Solving Problems in Soil Mechanics

Other titles in the Series

Solving Problems in Soil Mechanics

B. H. C. Sutton   BSc, MEng, CEng, MICE, MIStructE, MIOB, MBIM
Regional Co-ordinator,
Business & Technician Education Council

Longman
Scientific &
Technical

Copublished in the United States with
John Wiley & Sons, Inc., New York

Longman Scientific & Technical,
Longman Group UK Limited,
Longman House, Burnt Mill, Harlow,
Essex CM20 2JE, England
and Associated Companies throughout the world.

Copublished in the United States with
John Wiley & Sons, Inc., 605 Third Avenue, New York, NY 10158

First published 1986
Reprinted 1990

British Library Cataloguing in Publication Data
Sutton, B. H. C.
  Solving problems in soil mechanics.—(Solving problems)
  1. Soil mechanics—Problems, exercises, etc.
  I. Title
  624.1'5136'076   TA710

ISBN 0-582-98810-1

ISBN 0 470 206918 (USA only)

Produced by Longman Singapore Publishers (Pte) Ltd.
Printed in Singapore.

Contents

Preface

Soil mechanics is an engineering science which has continued to develop rapidly in recent years. The subject matter is introduced here by solving problems which highlight the engineering properties of soil and their implications for design.

Contributions to soil mechanics have been made by research workers in many countries and an attempt has been made to acknowledge all sources of material in the text.

The author is grateful to the Senate of the University of London, the Council of the Engineering Institutions, the Council of the Institution of Civil Engineers and the Scottish Technical Education Council for permission to use questions set in past examination papers.

B. H. C. Sutton   March 1985

1

Physical properties and classification of soil

The materials which constitute the earth's crust are divided, somewhat arbitrarily, by civil engineers into soil and rock. Soil is taken to refer to comparatively soft, loose and uncemented deposits which can be excavated by hand or using hand tools, while rock refers to hard, rigid bedrock and strongly cemented deposits.

Soil is formed by the disintegration of rock under the action of various forces of nature such as water, wind, frost, temperature changes and gravity. It may thus be considered to consist of a network of solid particles which enclose voids or pores. The voids may be filled with water or air or both. The soil is described as:

> *dry* if the voids are full of air
> *saturated* if the voids are full of water
> *partially saturated* if the voids contain both air and water.

A *phase* is one part of the soil system which is chemically and physically different from the other parts. Soil is thus a multi-phase material consisting of:

(1) solids (usually mineral particles);
(2) liquid (usually water);
(3) gas (usually air).

For the solution of engineering problems, it is often necessary to know the proportions by mass and volume of the various soil phases. In order to study these relationships, it is convenient to use a soil-phase diagram in which the various phases of the soil are separated as shown in Fig. 1.1.

The following terms are used in connection with soil-phase diagrams and in soil mechanics generally:

$$\text{void ratio} \qquad e = \frac{\text{total volume of voids}}{\text{total volume of solids}}$$

$$= \frac{V_V}{V_S}$$

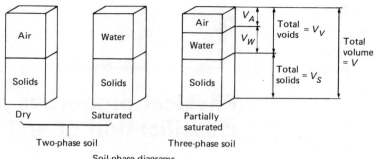

Dry   Saturated   Partially saturated

Two-phase soil   Three-phase soil

Soil-phase diagrams

Figure 1.1

porosity $\qquad n = \dfrac{\text{total volume of voids}}{\text{total volume of soil}}$

$$= \frac{V_V}{V} = \frac{V_V}{V_S + V_V} = \frac{e}{1 + e}$$

degree of saturation $S_r = \dfrac{\text{volume of water in voids}}{\text{total volume of voids}}$

$$= \frac{V_W}{V_V} \quad \text{(usually expressed as \%)}$$

Specific gravity
of solid particles $\quad G_s = \dfrac{\text{mass of solids}}{\text{mass of equal volume of water}}$

water (or moisture)
content $\quad w = \dfrac{\text{mass of water in voids}}{\text{mass of solids}} \times 100\%$

mass (or bulk) density
of soil $\quad \rho = \dfrac{\text{mass of soil}}{\text{volume of soil}}$

unit weight of soil $\quad \gamma = \dfrac{\text{weight of soil}}{\text{volume of soil}}$

In the SI system, the basic units are:

   length—measured in metres (m)
   mass—measured in kilograms (kg)
   time—measured in seconds (s).

Derived units are obtained by multiplication and division of unit values of these quantities. Thus the derived unit of force—the newton (N)—is obtained from:

$$1\,\text{N} = 1\,\text{kg} \times 1\,\text{m/s}^2$$

(since force = mass × acceleration).

In soil mechanics the unit weight (γ) is derived from the mass density (ρ) by:

$$\gamma = \rho \times g$$

where g is the acceleration due to gravity ($= 9.81$ m/s^2).
The most commonly used units are:

mass density—Mg/m^3 $\quad$ ($= $t/m^3)
unit weight—kN/m^3

Thus: mass density of water $\rho_w = 1\,000$ kg/m$^3 = 1.0$ Mg/m^3
unit weight of water $\gamma_w = 1.0 \times 9.81 = 9.81$ kN/m^3*

Worked examples

1.1 Water content of a soil sample

A moist sample of soil in a bottle had a mass of 25·24 g and the bottle, when empty, has a mass of 14·2 g. After drying in an oven for 24 hours, the bottle and soil sample had a mass of 21·62 g. Find the water content of the soil.

Solution This is the standard laboratory method of determining the water content of a soil sample. The oven temperature is maintained at about 105°C.

If $m_1 =$ mass of bottle
$m_2 =$ mass of bottle + wet soil
$m_3 =$ mass of bottle + dry soil

then

$$w = \frac{m_2 - m_3}{m_3 - m_1} \times 100\% = \frac{25.24 - 21.62}{21.62 - 14.20} \times 100 = \mathbf{49\%}$$

1.2 Specific gravity of a soil sample

A pycnometer having a mass of 620 g was used to determine the specific gravity of an oven dried sample of soil. If the mass of the soil sample was 980 g and the mass of the pycnometer with the sample and filled up with water was 2 112 g, determine the specific gravity of the soil particles. The mass of the pycnometer when filled with water only was 1495 g.

Solution A pycnometer is shown in Fig. 1.2. It is used in the determination of the specific gravity of the solid particles in the field. In the laboratory, a 1 litre jar is used together with a mechanical shaker. The calculations for either method are however the same.

* In view of the variable nature of soils, in the text the value of g has been taken as 10·0 m/s^2 and thus the unit weight of water is taken as 10 kN/m^3. Soil unit weights are derived by multiplying the mass density by 10.

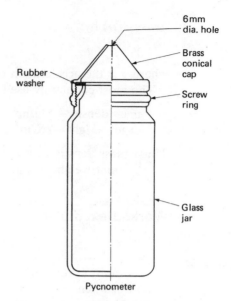

6mm
dia. hole

Brass
conical
cap

Rubber
washer

Screw
ring

Glass
jar

Pycnometer

Figure 1.2

The mass of the empty pycnometer (m_1) is found using a balance, a sample of the oven dried soil is placed inside and the combined mass (m_2) is found. Water is added to the soil which is agitated to remove all air pockets and when the pycnometer is full up, its mass (m_3) is measured. Finally the pycnometer is emptied, cleaned and filled with water and its new mass (m_4) found.

$$\therefore \text{ Mass of soil} = (m_2 - m_1) \text{ g.}$$

$$\text{Mass of water to fill jar} = (m_4 - m_1) \text{ g}$$
$$= \text{volume of water in m}^{-3}$$

$$\text{Mass of water to fill jar with soil in it} = (m_3 - m_2) \text{ g}$$
$$= \text{volume in m}^{-3}$$

$$\therefore G_s = \frac{\text{mass of soil}}{\text{volume of soil}} = \frac{(m_2 - m_1)}{(m_4 - m_1) - (m_3 - m_2)}$$

Substituting the given figures:

$$G_s = \frac{980}{(1495 - 620) - (2112 - 980 - 620)} = \textbf{2·7}$$

1.3 *In-situ* density of a soil sample

Describe the sand replacement method of determining the *in-situ* density of the soil.

The following results were obtained from a test:

Mass of soil extracted from hole	4·0 kg
Water content of soil	18%

Mass of dry soil to fill hole   3·1 kg
Mass of dry sand to fill container
    of volume 4·2 litres   5·8 kg

Calculate the wet and dry densities of the soil.

If the specific gravity of the particles is 2·68, find the degree of saturation of the soil.   (ICE)

Solution   The *in-situ* density of a soil is its bulk density in its natural undisturbed state in the ground. It is sometimes called the wet density.

The principle of determining the *in-situ* density of a soil is to remove a representative sample of the soil and measure its mass and volume. For cohesive soils, a core cutter of known volume is driven into the ground and extracted with a soil sample inside. For cohesionless soils, the sand replacement method is used.

A hole is excavated in the ground and all the soil removed is collected and its mass and moisture content found.

The volume of the hole excavated is measured by means of a sand pouring cylinder (Fig. 1.3).

The mass of the cylinder when filled with sand of a known density is measured. The cylinder is placed over the hole, the valve is opened

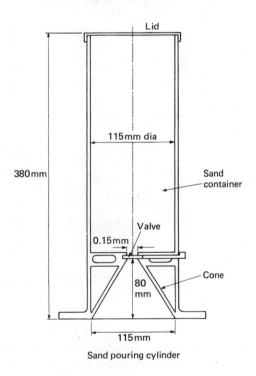

Sand pouring cylinder

Figure 1.3

and sand falls from the container into the hole until the hole and cone are full. The mass of the partially empty container is found using a balance so that the amount of sand poured out is known. The container is calibrated to find how much sand is required to fill the conical section and hence the mass of sand to fill the hole can be found. The density of the soil is then found by calculation as shown below.

$$\text{Density of sand} = \frac{\text{mass of sand}}{\text{volume of sand}} = \frac{5 \cdot 8}{4 \cdot 2} = 1 \cdot 38 \, \text{Mg/m}^3$$

$$\text{Volume of hole} = \frac{\text{mass of sand to fill hole}}{\text{density of sand}}$$

$$= \frac{3 \cdot 1 \times 10^{-3}}{1 \cdot 38} = 2 \cdot 25 \times 10^{-3} \, \text{m}^3$$

$$\text{Wet density of soil } \rho = \frac{\text{mass of soil excavated}}{\text{volume of hole}}$$

$$= \frac{4 \cdot 0 \times 10^{-3}}{2 \cdot 25 \times 10^{-3}} = 1 \cdot 78 \, \text{Mg/m}^3$$

Considering the soil-phase diagram (Fig. 1.4)

$$w = \frac{M_W}{M_S} \times 100 = 18$$

$$\therefore \ M_W = 0 \cdot 18 M_S$$

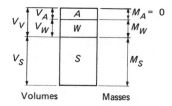

Figure 1.4

Also $M_W + M_S = 4 \cdot 0 \, \text{kg}$

$\therefore \ 0 \cdot 18 M_S + M_S = 4 \cdot 0 \, \text{kg}$

$\therefore \ M_S = 3 \cdot 39 \, \text{kg}$

$$\therefore \text{ dry density } \rho_d = \frac{3 \cdot 39 \times 10^{-3}}{2 \cdot 25 \times 10^{-3}} = 1 \cdot 51 \, \text{Mg/m}^3$$

$$\therefore \ V_S = \frac{3 \cdot 39}{1 \, 000 \times 2 \cdot 68} = 1 \cdot 265 \times 10^{-3} \, \text{m}^3$$

$$M_W = w \times M_S = 0 \cdot 18 \times 3 \cdot 39 = 0 \cdot 61 \, \text{kg}$$

$$\therefore \ V_W = \frac{0 \cdot 61}{1 \, 000} = 0 \cdot 61 \times 10^{-3} \, \text{m}^3$$

$$\text{Total volume} = 2\cdot25 \times 10^{-3} = V_A + V_W + V_S$$
$$\therefore V_A = (2\cdot25 - 0\cdot61 - 1\cdot265) \times 10^{-3}$$
$$= 0\cdot375 \times 10^{-3} \text{ m}^3$$
$$\therefore V_V = (0\cdot375 + 0\cdot61)10^{-3} = 0\cdot985 \times 10^{-3} \text{ m}^3$$
$$\therefore S_r = \frac{V_W}{V_V} = \frac{0\cdot61 \times 10^{-3}}{0\cdot985 \times 10^{-3}} = \mathbf{0\cdot62}$$

1.4 Void ratio of a soil sample

A sample of soil has a water content of 27 per cent and a bulk density of $1\cdot97 \text{ Mg/m}^3$. Determine the dry density and the void ratio of the soil, and the specific gravity of the particles.

What would be the bulk density of a sample of this soil with the same void ratio, but only 90 per cent saturated? (ICE)

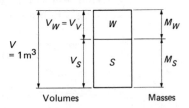

Figure 1.5

Solution If the bulk density of the soil is known, it is convenient to consider 1 m³ of the soil and draw the appropriate soil-phase diagram (Fig. 1.5).

$$w = \frac{M_W}{M_S} \times 100 = 27 \quad \therefore M_W = 0\cdot27 M_S$$

$$\rho = M_W + M_S = 1\cdot97 \text{ Mg/m}^3 \quad \therefore 0\cdot27 M_S + M_S = 1\cdot97$$

$$\therefore M_S = 1\cdot55 \text{ Mg and since 1 m}^3 \text{ of soil is being considered,}$$

$$\therefore \text{ dry density } \rho_d = \mathbf{1\cdot55 \text{ Mg/m}^3}$$

$$\therefore M_W = 0\cdot27 \times 1\cdot55 = 0\cdot42 \text{ Mg}$$

$$V_W = \frac{M_W}{\rho_w} = \frac{0\cdot42}{1\cdot0} = 0\cdot42 \text{ m}^3$$

$$\therefore V_S = 1 - 0\cdot42 = 0\cdot58 \text{ m}^3$$

$$e = \frac{V_V}{V_S} = \frac{0\cdot42}{0\cdot58} = \mathbf{0\cdot724}$$

$$M_S = V_S \times G_s \times \rho_w$$

$$\therefore G_s = \frac{1\cdot55}{0\cdot58 \times 1\cdot0} = \mathbf{2\cdot68}$$

If the voids are 90% saturated

$$M_W = 0{\cdot}42 \times 0{\cdot}90 = 0{\cdot}38 \text{ Mg}$$
$$M_S = 1{\cdot}55 \text{ Mg}$$

$\therefore M_W + M_S + (M_A = 0) = 1{\cdot}93 \text{ Mg}$
$\therefore \rho = \mathbf{1{\cdot}93 \ Mg/m^3}$ since the volume of the soil-phase diagram $= 1 \text{ m}^3$.

1.5 Degree of saturation of a soil sample

Derive an expression for the bulk density of partially saturated soil in terms of the specific gravity of the particles G_s, the void ratio e, the degree of saturation S_r and the density of water ρ_w.

In a sample of clay the void ratio is $0{\cdot}73$ and the specific gravity of the particles is $2{\cdot}7$. If the voids are 92 per cent saturated, find the bulk density, the dry density and the percentage water content.

What would be the water content for complete saturation, the void ratio being the same?   (ICE)

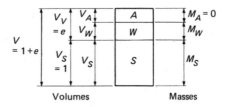

Figure 1.6

Solution In order to determine these relationships a soil-phase diagram is used (Fig. 1.6) in which the volume of the solid particles is 1. The volume of the voids will then be e by definition.

Volume of solids $V_S = 1$

Mass of solids $M_S = V_S . G_s . \rho_w = G_s . \rho_w$

Volume of voids $V_V = e$

Degree of saturation of voids $S_r = \dfrac{V_W}{V_V}$

$\therefore V_W = V_V . S_r = e . S_r$

Mass of voids $M_V = V_W . \rho_w = e . S_r . \rho_w$

$\therefore$ Partially saturated bulk density

$$\rho = \frac{M_S + M_V}{V_S + V_V} = \frac{G_s . \rho_w + e . S_r . \rho_w}{1 + e}$$

$$\therefore \rho = \frac{G_s + e . S_r}{1 + e} . \rho_w \tag{1.1}$$

Substituting given values

$$\rho = \frac{2 \cdot 7 + 0 \cdot 73 \times 0 \cdot 92}{1 + 0 \cdot 73} \times 1 \cdot 0 = \mathbf{1 \cdot 95\ Mg/m^3}$$

$$\rho_d = \frac{M_S}{V} = \frac{2 \cdot 7 \times 1 \cdot 0}{1 \cdot 73} = \mathbf{1 \cdot 56\ Mg/m^3}$$

$$w = \frac{M_W}{M_S} = \frac{0 \cdot 73 \times 0 \cdot 92 \times 1 \cdot 0}{2 \cdot 7 \times 1 \cdot 0} \times 100 = \mathbf{24 \cdot 9\%}$$

For complete saturation the voids are full of water.

$$\therefore S_r = 1$$

$$\therefore w = \frac{0 \cdot 73 \times 1 \times 1 \cdot 0}{2 \cdot 71 \times 1 \cdot 0} \times 100 = \mathbf{26 \cdot 9\%}$$

1.6 Formulae for density of a soil sample

Express
(a) the bulk density,
(b) the saturated density,
(c) the dry density,
(d) the submerged density,
of a soil in terms of the specific gravity of the particles G_s, the void ratio e, the degree of saturation of the voids S_r and the density of water ρ_w.

Solution The expression for the bulk density of a soil was developed in solution 1.5.

(a) $$\rho = \frac{G_s + e \cdot S_r}{1 + e} \cdot \rho_w \tag{1.1}$$

(b) For a saturated soil $S_r = 1$ $\quad \therefore \rho_{sat} = \dfrac{G_s + e}{1 + e} \cdot \rho_w \tag{1.2}$

(c) For a dry soil $S_r = 0$ $\quad \therefore \rho_d = \dfrac{G_s}{1 + e} \cdot \rho_w \tag{1.3}$

(d) For a submerged soil, i.e. a soil that is below water level, the density is reduced because of the buoyant effect that the water has on the solid particles.

$$\therefore \rho' = \rho_{sat} - \rho_w = \frac{G_s + e}{1 + e} \cdot \rho_w - \rho_w$$

$$= \frac{G_s - 1}{1 + e} \cdot \rho_w \tag{1.4}$$

Soil classification

A *class* is a group of soils which resemble one another in a few important characteristics. The process of classification, therefore, is the placing of a soil sample in its particular group or class and is of fundamental importance since it enables the soil to be described in terms which are internationally understood.

There are so many different properties of soil which are of interest to engineers and so many different combinations of these properties in any soil that a universal classification system is impractical. Thus all soil classification systems have been devised for specific purposes (e.g. pavement design) and this must be borne in mind when using a particular system.

All systems however are based on the particle sizes found within the soil mass and most systems recognize three main types of soil:

(1) coarse grained or cohesionless soil,
(2) fine grained or cohesive soil,
(3) organic soil.

The term *grain* is applied to the individual mineral particles in the soil.

For coarse grained soil, the most commonly used test is a mechanical analysis of the particle sizes since the size of the particles and the proportions of the different sizes have an important effect on the behaviour of the soil. In the case of fine grained soils, the consistency (i.e. the tendency of the particles to stick together) and the plasticity (i.e. the ability to deform without rupture) are of more significance than the grain size. These properties are affected by the water content and the commonly used tests are devised to enable the behaviour of the soil at various moisture contents to be studied.

1.7 Classification of a soil sample

(a) Explain briefly the object of classifying soil for engineering purposes.

(b) Discuss any method of classification known to you and state the physical properties and factors which are considered in that particular classification. (UL)

Solution (a) The purpose of classification is to obtain a consistent and internationally recognized description of the soil sample. This facilitates the interchange of general information about similar soils and forms a basis for decisions on further tests required for the solution of a particular engineering problem.

(b) Most classification methods have evolved from the Casagrande system in which coarse grained soils are classified on the basis of the size and distribution of the particles and fine grained soils on the basis of their plasticity using a chart.

For descriptive purposes, soils are designated by two letters. The first indicates the main soil type and the second denotes a qualifying sub-division to give a more accurate description. The most commonly used letters are shown on Table 1.1.

Table 1.1

Coarse grained soil	*Main terms*	**Gravel**	G
		Sand	S
	Qualifying terms	Well graded	W
		Poorly graded	P
		Uniformly graded	P_u
		Gap graded	P_g
Fine grained soil	*Main terms*	**Fine soil**	F
		Silt	M
		Clay	C
	Qualifying terms	Low plasticity	L
		Intermediate plasticity	I
		High plasticity	H
		Very high plasticity	V
		Extremely high plasticity	E
Organic soil		**Peat**	P_t
		Organic	O*

* May be applied to any soil containing significant amounts of organic material.

A coarse grained soil is one in which less than 35% of the material is finer than 0·06 mm. A fine grained soil contains more than 35% of material finer than 0·06 mm. Both types are further sub-divided on the basis of grain size as shown on Fig. 1.7.

If a sample of coarse grained soil is passed through a series of sieves, the amount retained on each weighed and the results plotted as shown on Fig. 1.8 the grading of the soil can be seen.

Fine grained soils are described by reference to their position on the plasticity chart shown on Fig. 1.9.

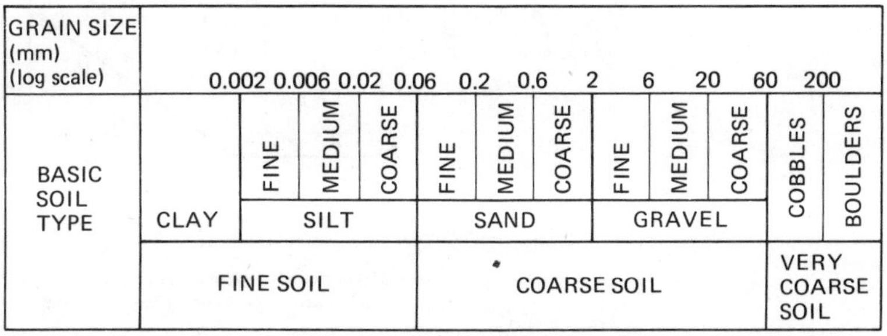

Figure 1.7

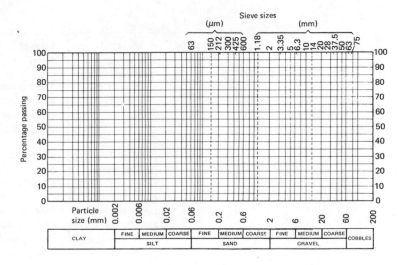

Figure 1.8

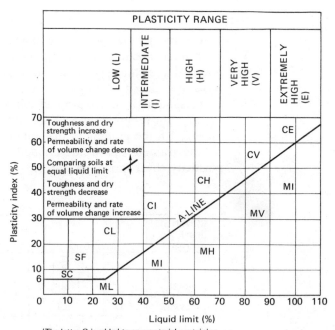

Figure 1.9

1.8 Liquid limit, plastic limit and plasticity index of a soil sample

Explain the terms *liquid limit, plastic limit* and *plasticity index,* and describe how they are measured.

The following results were obtained from a liquid limit test:

Number of taps	6	8	12	26	28	31
Water content (%)	53·4	52·2	48·3	40·0	38·8	37·1

Find the liquid limit.
Find also the plasticity index if the plastic limit was 18%.
What is the classification of this soil?

Solution When a fine grained soil is deposited from suspension in a liquid it passes through four states of *consistency* depending on the water content:

(1) liquid state;
(2) plastic state;
(3) semi-solid state;
(4) solid state.

The water contents at which the soil passes from one state to the next are called *consistency limits* and are expressed as *w*%.

The *liquid limit* (LL) is the water content at which the soil passes from the plastic to the liquid state, i.e. begins to behave like a viscous mud and flow under its own weight.

The *plastic limit* (PL) is the lowest water content at which the soil remains in a plastic state, i.e. when it is about to change from a plastic state to a crumbly semi-solid.

The *shrinkage limit* (SL) is the water content at which further loss of water in the soil will not cause further reduction in the volume of the soil, i.e. the water content required just to fill the voids of a sample which has been dried.

Various indices may be derived from these limits. The *plasticity index* (PI) is a measure of the range of water contents over which the soil remains in a plastic state.

$$PI = LL - PL$$

The *liquidity index* (I_L) is used to compare the *in-situ* water content with its plasticity.

$$I_L = \frac{w - PL}{PI}$$

One method of measuring the liquid limit is by means of the Casagrande apparatus shown in Fig. 1.10.

Wet soil is placed in the cup and divided into two halves by a grooving tool (Fig. 1.10). The cup is then tapped twice a second and

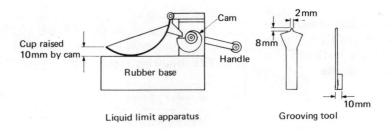

Liquid limit apparatus   Grooving tool

Figure 1.10

the number of taps required to bring the two halves together is recorded and the moisture content of the soil is found. The procedure is repeated for different water contents and a semi-logarithmic graph is plotted of water content against the number of taps. The water content of the soil corresponding to 25 taps on the graph is taken as the liquid limit of the soil.

The plastic limit is found by rolling a ball of wet soil between the palm of the hand and a glass plate to produce a thread 3 mm thick before the soil just begins to crumble. The water content of the soil in this state is taken as the plastic limit.

From the given figures the water content is plotted against the log of the number of taps (Fig. 1.11) and from this graph the liquid limit = **40%**.

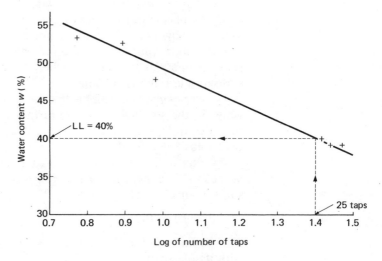

Figure 1.11

The plastic limit = 18%

∴ Plasticity index = 40 − 18 = **22%**

From Fig. 1.9 it can be seen that the soil may be classified as an inorganic clay of intermediate plasticity.

1.9 Use of cone penetrometer to find plastic limit of a soil sample

The following results were obtained from a plastic limit test on the same soil using the cone penetrometer apparatus. Determine the plastic limit of the soil.

Cone penetration (mm)	15·5	17·3	19·6	22·4	23·9
w (%)	47	51	56	63	66

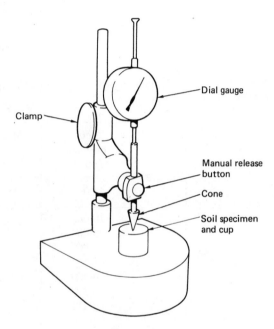

Figure 1.12

Solution A more recent method of finding the plastic limit of a soil, which is both easier and gives more reproducible results, is by means of a cone penetrometer shown in Fig. 1.12. A sample of at least 200 g of the soil is mixed with water on a glass sheet to make a uniform paste. This is placed in a metal cup and the surface is struck off level. The cone is lowered on to the surface of the soil and the dial gauge read. The cone is released and its penetration into the soil is measured.

The test is repeated at the same water content and again with the same soil at increasing water contents. A sample of the soil is used each time to determine the moisture content at the time of the test.

The results are plotted on Fig. 1.13 and the liquid limit is taken as the moisture content which corresponds with a cone penetration of 20 mm.

In the example given, the liquid limit read from Fig. 1.13 is **57%**.

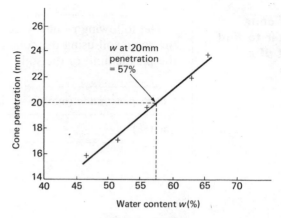

Figure 1.13

1.10 Particle size analysis of a soil sample

The results of a sieve analysis of two soil samples are as follows:

Sieve aperture (mm)	Mass of soil retained on sieve (g)	
	Soil A	Soil B
37·50	0·0	
20·00	26·0	
10·00	31·0	
5·00	11·0	0·0
2·00	18·0	8·0
1·18	24·0	7·0
0·600	21·0	11·0
0·300	41·0	21·0
0·212	32·0	63·0
0·150	16·0	48·0
0·063	15·0	14·0

A sedimentation test on the material passing the 63 μm sieve indicated that the samples contained:

Particle size (mm)	Mass (g)	
	A	B
0·06–0·02	8	2
0·02–0·006	4	1
0·006–0·002	2	0
less than 0·002	1	0

Plot the standard grading curves for the two soils and determine the effective size D_{10}, the coefficient of uniformity C_u and the coefficient of curvature C_c. Comment briefly on the results.
(HNC)

Solution The mechanical analysis of a soil is performed by passing a sample of oven dried soil through a series of graded sieves. The amount retained on each sieve is weighed. For the very fine material passing a 63 μm sieve, a sedimentation test is performed. The results obtained are plotted on a standard semi-logarithmic chart.

The total weight of the sample is found and then the percentage of the total weight retained on each sieve is calculated. From these figures the percentage of the sample passing through each sieve can be calculated by subtraction from 100%. This value is plotted against the corresponding sieve size or smallest particle size.

Sieve aperture (mm)	Soil A			Soil B		
	mass (g)	% retained	% passing	mass (g)	% ret.	% pass.
37·50	0	0	100			
20·00	26	10·4	89·6			
10·00	31	12·4	77·2			
5·00	11	4·4	72·8	0	0	100
2·00	18	7·2	65·6	8	4·6	95·4
1·18	24	9·6	56·0	7	4·0	91·4
0·600	21	8·4	47·6	11	6·3	85·1
0·300	41	16·4	31·2	21	12·0	73·1
0·212	32	12·8	18·4	63	36·0	37·1
0·150	16	6·4	12·0	48	27·4	9·7
0·063	15	6·0	6·0	14	8·0	1·7
0·02	8	3·2	2·8	2	1·1	0·6
0·006	4	1·6	1·2	1	0·6	—
0·002	2	0·8	0·4			
<0·002	1	0·4	—			
Total	250	100		175	100	

These results are plotted on Fig. 1.14.

The effective size of the sample, D_{10} = maximum size of particle in smallest 10% of sample.

The effective size represents an attempt to determine the diameter of spheres which would have the same effect as the given soil when used as a filtering medium.

The coefficient of uniformity $C_u = \dfrac{D_{60}}{D_{10}}$

where D_{60} = max. size of particle in smallest 60% of sample.

An important property of cohesionless soil is its degree of uniformity, i.e. whether the individual particles are mainly the same size or if there is a great range of sizes. The greater the value of C_u, the less uniform is the grading. In general a non-uniform (well-graded) soil has greater strength and stability than a uniform (poorly-graded) soil.

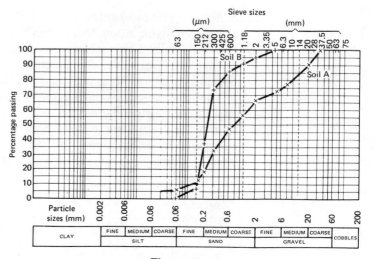

Figure 1.14

A well graded soil will have a value $C_u > 5$ and a poorly graded soil $C_u < 2$.

The coefficient of curvature $C_c = \dfrac{D_{30}^2}{D_{10}D_{60}}$.

This represents another way of comparing the shapes of the grading curves.

From Fig. 1.14 the following figures can be obtained:

Soil A $D_{10} = 0{\cdot}10$ mm $D_{30} = 0{\cdot}30$ mm $D_{60} = 1{\cdot}6$ mm
Soil B $D_{10} = 0{\cdot}15$ mm $D_{30} = 0{\cdot}19$ mm $D_{60} = 0{\cdot}26$ mm

$\therefore$ For soil A $C_u = \dfrac{1{\cdot}6}{0{\cdot}10} = 16$ $C_c = \dfrac{0{\cdot}30^2}{0{\cdot}10 \times 1{\cdot}6} = 0{\cdot}56$

For soil B $C_u = \dfrac{0{\cdot}26}{0{\cdot}15} = 1{\cdot}7$ $C_c = \dfrac{0{\cdot}19^2}{0{\cdot}15 \times 0{\cdot}26} = 0{\cdot}92$

Thus soil A is a well-graded (non-uniform) gravelly sand. Soil B is a poorly graded (uniform) slightly silty sand.

1.11 Comparison of properties of soil samples

The following Index Properties were determined for two soils X and Y:

Property	X	Y
Liquid limit	0·62	0·34
Plastic limit	0·26	0·19
Water content	38%	25%
Specific gravity of solids	2·72	2·67
Degree of saturation	1·00	1·00

Which of these soils: (*a*) contains more clay particles; (*b*) has a greater wet density; (*c*) has a greater dry density; (*d*) has a greater void ratio? Give reasons for your answers.　　(UL)

Solution　(*a*) The plasticity index is the range of moisture contents over which the soil remains plastic. The longer this range, the greater the proportion of clay particles.

For soil X, Plasticity index = Liquid limit − Plastic limit
$$= 0 \cdot 61 - 0 \cdot 26 = 0 \cdot 36$$

For soil Y, Plasticity index $= 0 \cdot 34 - 0 \cdot 19 = 0 \cdot 15$

∴ Soil X contains more clay particles.

(*b*) Consider 1 m^3 of soil. In both cases $S_r = 1$, i.e. the soil is fully saturated and the soil-phase diagram will be similar to Fig. 1.5.

Soil X

$w = \dfrac{M_w}{M_S} \times 100 = 38\%$

∴ $M_w = 0 \cdot 38 M_S$

∴ $V_w \cdot \rho_w = 0 \cdot 38 \cdot \rho_w \cdot G_s \cdot V_s$

∴ $V_w = 0 \cdot 38 \times 2 \cdot 72 \times V_s$

$= 1 \cdot 03 V_s$

But $V_w + V_s = 1$

∴ $2 \cdot 03 V_s = 1$

∴ $V_s = 0 \cdot 5$

∴ $V_w = 0 \cdot 5$

∴ $M_s = 0 \cdot 5 \times 2 \cdot 72 = 1 \cdot 36 \text{ Mg}$
$M_w = 0 \cdot 5 \times 1 \cdot 0 \ = 0 \cdot 50 \text{ Mg}$
∴ $\rho = \underline{1 \cdot 86 \text{ Mg/m}^3}$

Soil Y

$w = 25\%$

$M_w = 0 \cdot 25 M_S$

$V_w \cdot \rho w = 0 \cdot 25 \cdot \rho_w \cdot G_s \cdot V_s$

$V_w = 0 \cdot 25 \times 2 \cdot 67 \times V_s$

$= 0 \cdot 67 V_s$

∴ $1 \cdot 67 V_s = 1$

∴ $V_s = 0 \cdot 6$

∴ $V_w = 0 \cdot 4$

∴ $M_s = 0 \cdot 6 \times 2 \cdot 67 = 1 \cdot 60 \text{ Mg}$
$M_w = 0 \cdot 4 \times 1 \cdot 0 \ = 0 \cdot 40 \text{ Mg}$
∴ $\rho = \underline{2 \cdot 00 \text{ Mg/m}^3}$

Thus soil Y has the greater wet density.

(*c*) In this case M_S represents the dry density since 1 m^3 of soil is being considered.

∴ Soil Y has the greater dry density as well.

(*d*) Void ratio　$e = \dfrac{V_V}{V_S}$

Soil X　$e = \dfrac{0 \cdot 5}{0 \cdot 5} = 1$

Soil Y　$e = \dfrac{0 \cdot 4}{0 \cdot 6} = 0 \cdot 67$

Thus soil X has the greater void ratio.

Problems

1. Define the terms water content, porosity and void ratio as applied to a soil. Derive the relationship between the porosity and the void ratio.

In a test on a soil sample, the following results were obtained:

Total volume of soil in its natural state	645 ml
Total mass of soil after oven drying	1050 g
Mass of pycnometer	40 g
Mass of pycnometer full of water	674 g
Mass of pycnometer and dry soil	485 g
Mass of pycnometer and soil plus water	946 g

Determine the specific gravity of the soil particles and the void ratio of the soil. (HNC)

$$\left(n = \frac{e}{1+e}, \ 2\cdot57, \ 0\cdot58\right)$$

2. In order to determine the density of a clay soil, an undisturbed sample was taken in a sampling tube whose volume was $1\cdot6 \times 10^{-3} \, \text{m}^3$. The following data were obtained:

Mass of tube (empty)	1·86 kg
Mass of tube + clay sample	5·00 kg
Mass of tube + clay sample after oven drying	4·31 kg

Calculate the water content, the wet density and the dry density of the soil.

If the specific gravity of the soil particles was 2·69, find the void ratio and the degree of saturation of the clay. (ICE)

(28%, 1·96 Mg/m³, 1·53 Mg/m³; 0·75, 1·0)

3. A sample of clay soil of volume $1 \times 10^{-3} \, \text{m}^3$ and mass 1·762 kg, after being dried out in an oven had a mass of 1·368 kg. The specific gravity of the particles was 2·69. Find, for the sample in its original condition, (a) the dry density; (b) the water content; (c) the void ratio; and (d) the air voids (expressed as a percentage of the total voids).

Find also the water content and bulk density of a sample of this soil completely saturated, assuming that it has the same void ratio. (ICE)

[(a) 1·37 Mg/m³, (b) 29%, (c) 0·965, (d) 20%, 1·86 Mg/m³]

4. A sample of soil is prepared by mixing a quantity of dry soil (specific gravity of particles 2·7) with 10·5% by weight of water. Find the mass of this wet mixture which will be required to produce by static compaction, a cylindrical specimen 150 mm diameter by 125 mm deep with 5% air voids.

Find also the void ratio and the dry density. (ICE)

$(4.3 \text{ kg}, 0.375, 2.23 \text{ Mg/m}^3)$

5. A sample of dry soil is uniformly mixed with 16% by mass of water and compacted in a cylindrical mould. The volume of wet soil is $1 \times 10^{-3} \text{ m}^3$ and its mass 1·60 kg. If the specific gravity of the soil particles is 2·68, find the dry density, the void ratio and the degree of saturation.

Find also the water content necessary for complete saturation if the total volume remains the same. (ICE)

$(1.38 \text{ Mg/m}^3, 0.94, 0.46; 35\%)$

6. Describe a method of determining the *in-situ* density of a soil.

The *in-situ* density of a soil sample was found to be 2·10 Mg/m^3 and its water content was 15%. The specific gravity of the particles was 2·71. Calculate the dry density, the void ratio and the degree of saturation.

What would be the water content of this soil if completely saturated at the same void ratio? (ICE)

$(1.83 \text{ Mg/m}^3, 0.48, 0.85, 17.7\%)$

7. In a standard test the mass of soil contained in a mould (volume $1.6 \times 10^{-3} \text{ m}^3$) was 2·0 kg. Assuming the soil was 85% saturated, find the water content, the void ratio and the dry density of the soil, taking the specific gravity of the particles as 2·71. (ICE)

$(15\%, 0.49, 1.82 \text{ Mg/m}^3)$

8. An undisturbed specimen of clay was tested in the laboratory and the following results obtained:

Specific gravity: 2·70
Wet and oven-dried masses of specimen: 210 g and 125 g respectively.

Assuming the wet specimen to be (*a*) 100% saturated; (*b*) 75% saturated, determine (i) the total volume; (ii) the void ratio; (iii) the porosity of the specimen. (UL)

(100% Sat. vol. = (i) $131 \times 10^{-3} \text{ m}^3$, (ii) 1·85, (iii) 0·65;
75% Sat. vol. = (i) $159 \times 10^{-3} \text{ m}^3$, (ii) 2·46, (iii) 0·71)

9. Define the liquid and plastic limits of a soil, and explain their significance as a means of soil identification and classification. For a saturated soil derive the expressions for: (*a*) dry density; (*b*) bulk density; (*c*) submerged density.

An undisturbed sample of a saturated soil has a volume of $143 \times 10^3 \text{ mm}^3$ and a mass of 260 g. Determine its void ratio, water content and dry density.

A sample of a different soil was found to have a bulk density of $1.92\,\text{Mg/m}^3$ and a water content of 30%. Find the degree of saturation of the soil.

(Take the grain specific gravity $= 2.7$ for each soil.)　　(ICE)

$(1.07, 40\%, 1.30\,\text{Mg/m}^3; 98\%)$

10. In a test to determine the liquid limit of a silty clay the following results were recorded:

Test	W_1	W_2	W_3	Cone penetration (mm)
1	11·62	21·28	19·08	28·1
2	10·87	19·50	17·24	22·9
3	11·21	21·26	18·39	19·2
4	10·46	19·62	16·74	15·5

where $W_1 =$ mass of container (g)
　　　$W_2 =$ mass of container and wet soil (g)
　　　$W_3 =$ mass of container and soil after oven drying (g)

Determine the liquid limit for the soil. If the plastic limit for the soil was 22% and the natural water content 35% find the plasticity index and liquidity index.　　(HNC)

$$\left(37\%,\ 15\%,\ \text{LI} = \frac{w - \text{PL}}{\text{PI}} = 0.87\right)$$

11. The results of a sieve analysis on a soil sample were:

Retained on 0·60 mm　 8 g
Retained on 0·212 mm 26 g
Retained on 0·063 mm 43 g

and a sedimentation test indicated that the soil contained:

Dia. of particles 0·06 −0·02 mm　 43 g
　　　　　　　　0·02 −0·006 mm　 17 g
　　　　　　　　0·006−0·002 mm　 7 g
　　　　　　　　less than 0·002 mm 3 g

Plot the standard grading curve for the soil, determine the effective size and coefficient of uniformity for the soil and describe the soil.

(0·01, 10, well graded silty sand)

2

Permeability of soil and flow nets

The voids in the soil are not isolated cavities which store water like reservoirs but small irregular passages through which water will flow. Therefore, since it contains continuous voids, soil is said to be a *permeable* material.

The source of water in soil is mainly rainfall. Water below the surface of the ground is known as *subsurface water*. It can be divided into different zones:

(1) A *saturation zone*. Here the top surface of the water is at atmospheric pressure and is known as the *water table* or *phreatic surface*. Below this level, the soil will be saturated with water subject to hydrostatic pressure.

(2) An *aeration zone* which lies between the water table and the ground surface. It may be divided into three sub-zones. Immediately above the water table, the soil remains saturated with water due to capillary action which holds the water below atmospheric pressure. This is often referred to as the *capillary fringe*. Above it is a partially saturated sub-zone where water is held by surface tension and adsorption. The top sub-zone occurs when evaporation upwards is continuously taking place.

Worked examples

2.1 Effective and pore water pressures in a soil mass

(*a*) Define the terms *effective pressure* and *pore water pressure*.

(*b*) A layer of sand 4·5 m deep overlies a thick bed of clay. The water table is 2 m below the top of the sand. Above the water table, the sand has an average void ratio of 0·52 and an average degree of saturation of 0·37. The clay has a water content of 42%.

Calculate the total, effective and pore water pressures on a horizontal plane 9 m below the ground surface and draw pressure distribution diagrams down to this level.

Assume the grain specific gravity = 2·65 for both the sand and the clay. (HNC)

Solution (*a*) The total vertical pressure σ on any horizontal plane below the surface of a soil, unloaded except for its own weight, is equal to

$$\frac{\text{force exerted by mass of overlying soil}}{\text{area of plane}}.$$

It can be divided into two distinct components—the effective pressure σ' and the pore water pressure u.

The *effective pressure* is the pressure transmitted through the points of contact of the solid particles of the soil. It is known as the effective pressure since any change in σ' produces a change in all the other mechanical properties of the soil. (It is sometimes referred to as the intergranular pressure.)

The *pore water pressure* is the pressure exerted by the mass of water in the voids below the level of the water table. It is also referred to as the neutral pressure since a change in u has no measurable effect on the mechanical properties of the soil.

Under equilibrium conditions i.e. when there is no flow of water through the soil, the total vertical pressure σ on any horizontal plane is made up of two components:
(1) the effective pressure σ',
(2) the pore water pressure u.
The equation

$$\sigma = \sigma' + u \qquad\qquad (2.1)$$

is a fundamental one in soil mechanics.

At the water table $u = 0$. Above the water table, the water in the voids of the soil is held by surface tension forces between the soil particles. The effect of this is to increase the effective pressure due to the mass of water held in suspension.

(*b*) Above the water table, the sand is partially saturated.

$$e = 0.52, \ S_r = 0.37, \ G_s = 2.65$$

and from solution 1.6

$$\rho = \left(\frac{G_s + S_r \cdot e}{1 + e}\right)\rho_w$$

$$= \left(\frac{2.65 + 0.37 \times 0.52}{1 + 0.52}\right)1.0 = 1.87 \ \text{Mg/m}^3$$

$\therefore$ Unit weight $\gamma = 1.87 \times 10 = 18.7 \ \text{kN/m}^3$ (see footnote, p.3)

Below the water table, the sand is fully saturated and $S_r = 1$.

$$\therefore \rho_{\text{sat}} = \left(\frac{2.65 + 1.0 \times 0.52}{1 + 0.52}\right)1.0 = 2.09 \ \text{Mg/m}^3$$

$\therefore$ Unit weight $\gamma_{\text{sat}} = 2.09 \times 10 = 20.9 \ \text{kN/m}^3$

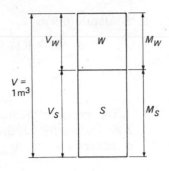

Figure 2.1

Clay:
Consider 1 m³ of soil (Fig. 2.1): $w = 42\%$, $G_s = 2.65$, $S_r = 1$.

$$\frac{M_W}{M_S} = 0.42 \quad \therefore \quad M_W = 0.42 M_S$$

$$V_W + V_s = 1 \qquad \therefore \quad \frac{M_W}{\rho_w} + \frac{M_S}{\rho_w \cdot G_s} = 1$$

$$\therefore \quad \frac{0.42 M_S}{1.0} + \frac{M_S}{1.0 \times 2.65} = 1$$

$$\therefore \quad M_S \qquad\qquad = 1.254\ \text{Mg}$$

$$\therefore \quad M_2 = 0.42 \times 1.254 = 0.526\ \text{Mg}$$

$$\therefore \quad \rho_{\text{sat}} \qquad\qquad = 1.780\ \text{Mg/m}^3$$

$$\therefore \quad \gamma_{\text{sat}} = 1.78 \times 10 \quad = 17.80\ \text{kN/m}^3$$

Consider a horizontal plane 1 m² in plan with a column of soil 1 m²
above it (Fig. 2.2(a)).
At ground level, pressure $\sigma =$ zero.
At water table,

$$\text{Total pressure } \sigma = \frac{\text{force due to weight of column of sand 2 m high}}{\text{area of plane}}$$

$$\therefore \quad \sigma = \frac{\gamma h}{b^2} = \frac{18.7 \times 2}{1 \times 1} = 37.4\ \text{kN/m}^2$$

Pore water pressure $u = 0\ \text{kN/m}^2$
$\therefore$ Effective pressure $\sigma' = \sigma - u = 37.4\ \text{kN/m}^2$

(From eqn (2.1))

At top of clay

$$\sigma = 37.4 + 20.9 \times 2.5 = 89.7\ \text{kN/m}^2$$

$$u = 10 \times 2.5 \qquad\qquad = 25.0\ \text{kN/m}^2$$

$$\therefore \quad \sigma' = \sigma - u \qquad\qquad = 64.7\ \text{kN/m}^2$$

At depth of 9 m

$$\sigma = 89 \cdot 7 + 17 \cdot 8 \times 4 \cdot 5 = 169 \cdot 8 \text{ kN/m}^2$$
$$u = 25 \cdot 0 + 10 \times 4 \cdot 5 = \underline{70 \cdot 0 \text{ kN/m}^2}$$
$$\therefore \ \sigma' = \sigma - u = 99 \cdot 8 \text{ kN/m}^2$$

The pressure diagrams representing these figures are shown in Fig. 2.2. They may be shown on separate diagrams (b), (c) and (d) or combined on one diagram (e).

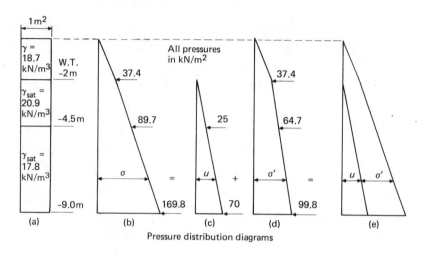

Pressure distribution diagrams

Figure 2.2

Seepage pressure

Any change from the equilibrium conditions will cause water to flow through the soil and this will alter the effective and pore water pressures.

2.2 Effect of seepage pressure on a soil mass

> Explain the meaning of the term *seepage pressure*. Show how the effective pressure is altered when water is flowing through the soil (a) vertically downwards; and (b) vertically upwards.
>
> (HNC)

Solution When water flows through a permeable soil it exerts a frictional drag force on the soil particles. The effect of this force per unit volume is known as the *seepage pressure*.

Figure 2.3 illustrates diagrammatically equilibrium conditions. The left hand vessel contains water and is connected to the right hand vessel containing soil and water. When the water level is the same in both vessels there will be no flow of water through the soil.

γ_{sat} = unit weight of the soil and γ_w = unit weight of water

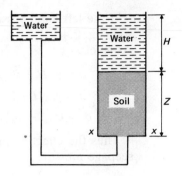

Figure 2.3

Then at level x–x

$$\sigma = \gamma_w H + \gamma_{sat} Z$$
$$u = \gamma_w(H + Z)$$
$$\therefore \ \sigma' = \sigma - u = (\gamma_{sat} - \gamma_w)Z \qquad (2.2)$$

(*a*) Flow downwards through the soil

If the left hand vessel is lowered, and the level of water in the right-hand vessel is maintained, water will flow downwards through the soil (Fig. 2.4(*a*)).

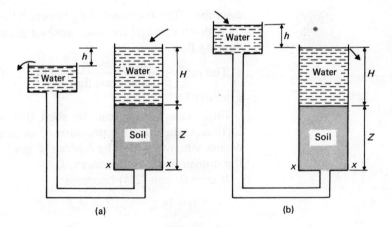

(a) (b)

Figure 2.4

At level x–x

$$\sigma = \gamma_w H + \gamma_{sat} Z$$
$$u = \gamma_w(H + Z - h)$$
$$\therefore \ \sigma' = (\gamma_{sat} - \gamma_w)Z + \gamma_2 h \qquad (2.3)$$

Thus the effective pressure is *increased* by $\gamma_w h$. This quantity $\gamma_w h$ is the seepage pressure exerted by the flowing water.

(*b*) Flow upwards through the soil

If the left hand vessel is raised water will flow upwards through the soil (fig. 2.4(b)).

At level $x-x$

$$\sigma = \gamma_w H + \gamma_{sat} Z$$

$$u = \gamma_w (H + Z + h)$$

$$\therefore \ \sigma' = (\gamma_{sat} - \gamma_w) Z - \gamma_w h \qquad (2.4)$$

Thus the effective pressure is *decreased* by $\gamma_w h$, the amount of the seepage pressure.

2.3 Critical hydraulic pressure in a soil mass

Excavation is being carried out in a soil with porosity $n = 0.35$ and grain specific gravity $G_s = 2.65$. Define what is meant by critical hydraulic gradient, and evaluate it for this soil.

A 1·25 m layer of the soil is subject to an upward seepage head of 1·85 m. What depth of coarse sand would be required above the soil to provide a factor of safety against 'piping'? (Assume that the coarse sand has the same porosity and grain specific gravity as the soil, and that there is negligible head loss in the sand). (ICE)

Solution The frictional drag between the solid particles and the water as it flows through the soil causes a gradual loss in the head of water causing flow.

The ratio $\dfrac{\text{head lost } h}{\text{distance travelled } Z}$ (Fig. 2.4(b)) is known as the hydraulic gradient i.

From eqn (2.4) it can be seen that as the seepage pressure $\gamma_w h$ increases, the effective pressure σ' decreases, until a critical condition occurs when $\sigma' = 0$. The hydraulic gradient for this case is known as the *critical hydraulic gradient*, i_c.

If $\sigma' = 0$, eqn (2.4) becomes:

$$0 = (\gamma_{sat} - \gamma_w) Z - \gamma_w h$$

$$\therefore \ \frac{h}{Z} = \frac{\gamma_{sat} - \gamma_w}{\gamma_w} = \frac{\rho_{sat} - \rho_w}{\rho_w} = i_c$$

From eqn (1.2)

$$\rho_{sat} = \frac{G_s + e}{1 + e} \cdot \rho_2$$

$$\therefore \ i_c = \frac{\dfrac{G_s + e}{1 + e} \cdot \rho_2 - \rho_2}{\rho_w} = \frac{G_s - 1}{1 + e} \qquad (2.5)$$

For the given soil

$$\text{Porosity } n = 0.35 = \frac{e}{1+e}$$

$$\therefore e = 0.54$$

$$\therefore i_c = \frac{2.65 - 1}{1 + 0.54} = 1.07$$

When the critical condition is reached ($\sigma' = 0$), the soil will become unstable. The water will force the overlying soil mass upwards and the surface will appear to 'boil' as the soil and water flow away together leaving a hole or 'pipe' in the soil. This phenomenon is known as 'piping' and is extremely serious.

Figure 2.5 illustrates the given conditions.

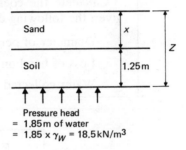

Sand

Soil

x

z

1.25 m

Pressure head
= 1.85 m of water
= 1.85 × γ_W = 18.5 kN/m³

Figure 2.5

The upward seepage head is 1.85 m of water

$$= 1.85 \times \gamma_w = 1.85 \times 10 = 18.5 \text{ kN/m}^2$$

The saturated density of the soil $\rho_{\text{sat}} = \dfrac{2.65 + 0.54}{1 + 0.54} \times 1.0 = 2.07 \text{ Mg/m}^3$

$$\therefore \text{ Unit weight } \gamma_{\text{sat}} = 2.07 \times 10 = 20.7 \text{ kN/m}^3$$

To provide a factor of safety against piping of 2, the effective pressure at the bottom of the soil must be twice the upward seepage pressure.

$$\therefore \ \sigma' = (\gamma_{\text{sat}} - \gamma_w)Z = 2\gamma_w h$$

$$(20.7 - 10)Z = 2 \times 10 \times 1.85$$

$$\therefore \ Z = \frac{37}{10.7} = 3.46 \text{ m}$$

$\therefore$ Depth of coarse sand required

$$x = 3.46 - 1.25 = \textbf{2.21 m}$$

Darcy's law

The flow of water through the soil is assumed to be 'laminar', that is each particle of water travels along a definite path which does not intersect the path of any other particle.

Darcy found that the flow of water through a soil in mm/s is proportional to the hydraulic gradient i. Thus Darcy's Law states that the velocity of flow through a column of saturated soil is proportional to the hydraulic gradient,

i.e. $v \propto i$

$\therefore$ $v = k \cdot i$ where k is known as the *coefficient of permeability* of the soil.

2.4 Permeability of a soil sample using a constant head permeameter

Describe with a neat sketch the constant head permeameter and assuming Darcy's law, derive an expression for the coefficient of permeability.

Calculate the coefficient of permeability of a sample of sand given the following data:

Diameter of permeameter	75 mm
Loss of head on a 200 mm length	83·2 mm
Water collected in 1 min	66·8 ml. (ICE)

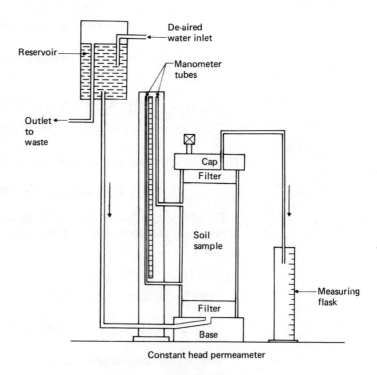

Constant head permeameter

Figure 2.6

Solution　A constant head permeameter is a device for measuring the permeability of a coarse grained soil (Fig. 2.6).

The permeameter consists essentially of a cylinder about 100 mm dia and 300 mm long fitted with a brass base and flanged brass head clamped together by vertical tie rods (not shown). A balancing reservoir is connected to the base plate and another connection from the head is taken to a measuring cylinder. Two connections from the cylinder are connected to manometer tubes.

The soil sample is placed in the cylinder between two filters and the test is performed by allowing water to flow through the soil sample at a head controlled and kept constant by the overflow in the balancing reservoir. When steady conditions are obtained, the time taken to collect a certain amount of water in the measuring cylinder is recorded. At the same time the difference in the level of the water in the two manometer tubes is recorded. This represents the loss of head as the water flows through a length of soil equal to the distance between the connections. The internal diameter of the cylinder is also measured.

From Darcy's law if k = coefficient of permeability

$$\text{seepage velocity } v = k \cdot i = k \cdot \frac{h}{l}$$

$$\text{Discharge in unit time } q = \frac{Q}{t} = \text{area} \times \text{velocity}$$

$$= \frac{\pi}{4} d^2 \times k \frac{h}{l}$$

$$\therefore \ k = \frac{4ql}{\pi d^2 h} \tag{2.6}$$

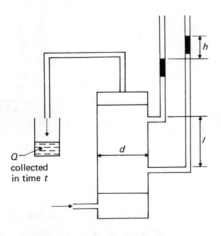

Figure 2.7

From data given

$$d = 75 \text{ mm}; \; l = 200 \text{ mm}; \; h = 83 \cdot 2 \text{ mm}$$

$$q = \frac{66 \cdot 8 \text{ ml}}{1 \text{ min}} \left[\frac{\text{min}}{60 \text{ s}} \times \frac{10^3 \text{ mm}^3}{\text{ml}} \right] = 1 \cdot 113 \times 10^3 \text{ mm}^3/\text{s}$$

$$\therefore \; k = \frac{4 \times 1 \cdot 113 \times 10^3 \times 200}{\pi \times 75^2 \times 83 \cdot 2} = 0 \cdot 606 \text{ mm/s}$$

2.5 Permeability of a soil sample using a variable head permeameter

Describe briefly with a sketch the variable head permeameter.

A permeameter of this type has a diameter of 75 mm and the length of the soil sample is 150 mm. The diameter of the standpipe is 15 mm. During the test the head decreased from 1 300 mm to 800 mm in 135 s. Calculate the coefficient of permeability of the soil in mm/s. Prove the formula used assuming Darcy's law. (ICE)

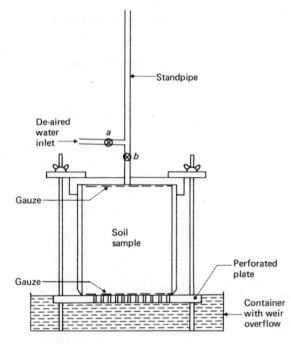

Falling head permeameter

Figure 2.8

Solution The variable (or falling) head permeameter is used to determine the permeability of very fine sand and silt (Fig. 2.8).

It consists essentially of a cylinder containing the soil sample and fitted with a cap and standpipe at the top and a perforated plate at the bottom.

Stopcock b is closed and a opened to fill the standpipe with water. Cock a is then closed and b opened and the time for the water to fall from one recorded level to another in the standpipe is noted. The length of the cylinder and the diameters of the cylinder and standpipe are also recorded.

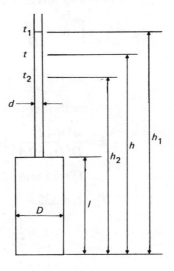

Figure 2.9

Figure 2.9 shows a diagram of the arrangement. Let the head above outlet at any time $t = h$. In a short time dt the head will fall $-dh$ (the $-$ ve sign indicates a loss of head).

$$\therefore \text{ Volume of water flowing out of standpipe } dv = -\frac{\pi}{4} d^2 \times dh$$

$$(2.7)$$

This must pass through the soil sample and since also

$$dv = \text{area} \times \text{velocity} \times \text{time}$$

$$\therefore \ dv = \frac{\pi}{4} D^2 \times v \times dt$$

From Darcy's Law

$$\text{seepage velocity } v = k \cdot i = \frac{kh}{l}$$

$$\therefore \ dv = \frac{\pi}{4} D^2 \times \frac{kh}{l} \times dt \qquad\qquad (2.8)$$

Equating eqns (2.7) and (2.8)

$$-\frac{\pi}{4}d^2 \times dh = \frac{\pi}{4}D^2 \times \frac{kh}{l} \times dt$$

$$\therefore \quad dt = \frac{-d^2}{D^2} \times \frac{l}{k} \times \frac{dh}{h}$$

The time taken for the water in the standpipe to fall from h_1 to h_2 may be found by integrating this expression between the appropriate limits.

$$\int_{t_1}^{t_2} dt = \frac{-d^2}{D^2} \cdot \frac{l}{k} \int_{h_1}^{h_2} \frac{dh}{h}$$

$$\therefore \quad t_2 - t_1 = \frac{d^2}{D^2} \cdot \frac{l}{k} \cdot \log_e \frac{h_1}{h_2}$$

$\therefore$ coefficient of permeability

$$k = \frac{d^2}{D^2} \frac{l}{(t_2 - t_1)} \cdot \log_e \frac{h_1}{h_2} \qquad (2.9)$$

$$D = 75 \text{ mm} \qquad d = 15 \text{ mm} \qquad h_1 = 1\,300 \text{ mm} \qquad h_2 = 800 \text{ mm}$$

$$t_2 - t_1 = 135 \text{ s} \qquad l = 150 \text{ mm}$$

$$\therefore \quad k = \frac{15^2}{75^2} \times \frac{150}{135} \times 2 \cdot 3 \log_{10} \frac{1\,300}{800} = 0 \cdot 0215 \text{ mm/s}.$$

2.6 Average permeability of two layers of soil

The data given below relate to two falling head permeability tests performed on different soils:

Standpipe area	400 mm^2
Permeameter sample area	2 800 mm^2
Permeameter sample height	50 mm
Initial water head in standpipe	1 000 mm
Final water head in standpipe	200 mm

Time for decreasing	soil 1	500
the water head	soil 2	15

Determine the coefficient of permeability of each of these soils in mm/s.

If these two soils form adjacent layers each 1·5 m thick, calculate the average permeability in directions parallel and orthogonal to the layers.

List two possible causes of error in falling head permeability tests.

(CEI)

Solution For falling head permeability (eqn (2.9))

$$k = \frac{d^2}{D^2} \cdot \frac{l}{(t_2 - t_1)} \cdot \log_e \frac{h_1}{h_2}$$

or in terms of area a of standpipe and area A of sample

$$k = \frac{a}{A} \cdot \frac{l}{(t_2 - t_1)} \cdot \log_e \frac{h_1}{h_2}$$

Then for soil 1,

$$k_1 = \frac{400 \times 50}{2\,800 \times 500} \times \log_e \frac{1\,000}{200} = 23 \times 10^{-3}\ \text{mm/s}$$

For soil 2

$$k_2 = \frac{400 \times 50}{2\,800 \times 15} \times \log_e \frac{1\,000}{200} = 766 \times 10^{-3}\ \text{mm/s}$$

Consider water flowing vertically downwards through two layers of soil of unit cross-sectional area but with different coefficients of permeability (Fig. 2.10).

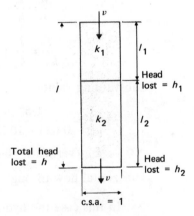

Figure 2.10

The quantity of water flowing through each section in unit time will be the same (by the principle of continuity of flow).

The quantity flowing in unit time $q = $ area $\times$ velocity and since the two cross-sectional areas are the same the velocity of flow v in each will be the same.

From Darcy's law $v = k \cdot i$

for the 1st layer $v = k_1 \cdot \dfrac{h_1}{l_1}$ where $h_1 = $ head lost in layer

$$\therefore \frac{h_1}{v} = \frac{l_1}{k_1} \tag{1}$$

for the 2nd layer $v = k_2 \cdot \dfrac{h_2}{l_2}$ where $h_2 =$ head lost in layer

$$\therefore \quad \frac{h_2}{v} = \frac{l_2}{k_2} \qquad (2)$$

If k_{av} is the average coefficient of permeability for the two layers

$$\therefore \quad v = k_{av}\left[\frac{h_1}{l_1} + \frac{h_2}{l_2}\right] = k_{av}\frac{h_2}{l}$$

$$\therefore \quad \frac{h}{v} = \frac{l}{k_{av}} \qquad (3)$$

Adding (1) and (2)

$$\frac{h_1}{v} + \frac{h_2}{v} = \frac{l_1}{k_1} + \frac{l_2}{k_2}$$

$$\therefore \quad \frac{1}{v}[h_1 + h_2] = \frac{l_1}{k_1} + \frac{l_2}{k_2}$$

$$\therefore \quad \frac{h}{v} = \frac{l_1}{k_1} + \frac{l_2}{k_2} \text{ and substituting from (3)}$$

$$\therefore \quad \frac{l}{k_{av}} = \frac{l_1}{k_1} + \frac{l_2}{k_2} \qquad (2.10)$$

Substituting values

$$\frac{3}{k_{av}} = \frac{1 \cdot 5}{0 \cdot 023 \times 10^{-3}} + \frac{1 \cdot 5}{0 \cdot 766 \times 10^{-3}}$$

$$\therefore \quad k_{av} \text{ (vertical)} = 0 \cdot 0446 \times 10^{-3} \text{ m/s}$$

Horizontal flow through two layers

In this case the hydraulic gradient is uniform $= i$

$$\text{Velocity } v = k_{av} \cdot i = \frac{1}{l}[v_1 l_1 + v_2 l_2]$$

$$= \frac{1}{l}[k_1 i \cdot l_1 + k_2 i \cdot l_2]$$

$$\therefore \quad k_{av} = \frac{k_1 l_1 + k_2 l_2}{l} \qquad (2.11)$$

$$= \frac{0 \cdot 023 \times 10^{-3} \times 1 \cdot 5 + 0 \cdot 766 \times 10^{-3} \times 1 \cdot 5}{3 \cdot 0}$$

$$\therefore \quad k_{av} \text{ (parallel)} = 0 \cdot 395 \times 10^{-3} \text{ m/s}$$

Two possible experimental errors which may occur in a falling head permeability test are:

(1) air bubbles may form in the sample;
(2) a filter of fine material may form on the surface of the sample.

2.7 *In-situ* permeability of a solid mass

A stratum of sandy soil overlies a horizontal bed of impermeable material, the surface of which is also horizontal. In order to determine the *in-situ* permeability of the soil, a test well was driven to the bottom of the stratum. Two observation boreholes were made at distances of 12·5 and 25 m respectively from the test well. Water was pumped from the test well at the rate of 3×10^{-3} m³/s until the water level became steady. The heights of the water in the two boreholes were then found to be 4·25 m and 6·50 m above the impermeable bed. Find the value, expressed in m³ per day, of the coefficient of permeability of the sandy soil, deriving any formula used. (ICE)

Solution When water is pumped from the ground it has the effect of lowering the water table over a large area. This is shown diagrammatically in Fig. 2.11.

Consider the flow inwards through the boundary of a cylindrical

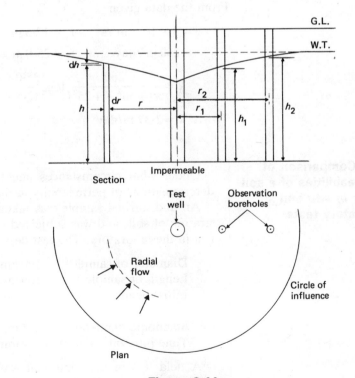

Figure 2.11

section of radius r where the water table is at a height h above the impermeable surface.

The quantity flowing in unit time $= \dfrac{Q}{t} = q = $ area $\times$ velocity and assuming Darcy's law $v = k \cdot i$

$$q = 2\pi r \cdot h \times k \cdot i = 2\pi r \cdot h \times k \cdot \frac{dh}{dr}$$

$$\therefore \quad \frac{dr}{r} = \frac{2\pi k}{q} \cdot h \cdot dh$$

Integrating between the appropriate limits for the two observation boreholes

$$\int_{r_1}^{r_2} \frac{dr}{r} = \frac{2\pi k}{q} \int_{h_1}^{h_2} h \cdot dh$$

$$\therefore \quad \log_e \frac{r_2}{r_1} = \frac{2\pi k}{q} \left(\frac{h_2^2}{2} - \frac{h_1^2}{2} \right)$$

$\therefore$ Coefficient of permeability

$$k = \frac{q}{\pi(h_2^2 - h_1^2)} \log_e \frac{r_2}{r_1} \tag{2.12}$$

From the data given

$$q = 3 \times 10^{-3} \, \text{m}^3/\text{s} \qquad r_1 = 12 \cdot 5 \, \text{m} \qquad h_1 = 4 \cdot 25 \, \text{m}$$
$$= 260 \, \text{m}^3/\text{day} \qquad r_2 = 25 \cdot 0 \, \text{m} \qquad h_2 = 6.50 \, \text{m}$$

$$\therefore \quad k = \frac{260}{\pi(6 \cdot 50^2 - 4 \cdot 25^2)} \log_e \frac{25 \cdot 0}{12 \cdot 5}$$

$$\therefore \quad k = \textbf{2·37 m/day}$$

2.8 Comparison of permeabilities of a soil using *in situ* and laboratory tests

Under what circumstances might it be justified to carry out determinations of permeability in the field?

An undisturbed sample was taken from a borehole made in a stratum of soil, and was subjected to a falling head permeability test in the laboratory. The test details were:

Diameter of sample	100 mm
Length of sample	100 mm
Initial head	450 mm
Final head	380 mh
Standpipe diameter	3 mm
Time interval	4 min

A field test in the same soil was made, in which water was pumped from a well which penetrated the full depth of the soil stratum. Under equilibrium conditions water was pumped from the well at the rate of $8.5\,m^3/day$. The observed height of the water level above the horizontal base of the stratum (which rested on impervious rock) was $4.5\,m$ and $5.5\,m$ in boreholes made at distances of $15\,m$ and $30\,m$ from the well, respectively.

Calculate the coefficient of permeability (in m/day) from each test. Comment on the results. (ICE)

Solution The permeability of a soil in a horizontal direction may be very different from the permeability in a vertical direction. There may also be layers of soil with different coefficients of permeability. In these circumstances a field test to find the equivalent coefficient of permeability of the soil would be more likely to give reliable results than those obtained in the laboratory.

Falling head permeameter test (see solution 2.5).

$$k = \frac{d^2}{D^2} \cdot \frac{l}{(t_2 - t_1)} \cdot \log_e \frac{h_1}{h_2}$$

$$\therefore\ k = \frac{3^2}{100^2} \times \frac{100}{4} \times \log_e \frac{450\,mm}{380\,min} \left[\frac{m}{10^3\,mm} \cdot \frac{60 \times 24\,min}{day} \right]$$

$$= 5.43 \times 10^{-3}\,\textbf{m/day}$$

Field pumping test (see solution 2.8).

$$k = \frac{q}{\pi(h_2^2 - h_1^2)} \log_e \frac{r_2}{r_1}$$

$$\therefore\ k = \frac{8.5}{\pi(5.5^2 - 4.5^2)} \log_e \frac{30}{15} = 187 \times 10^{-3}\,\textbf{m/day}$$

The big difference in the results of the two tests may be due to a difference in the vertical and horizontal coefficients of permeability of the soil. The falling head test measured the value of k in a vertical direction whereas the pumping tends to measure the permeability in a horizontal direction.

2.9 Permeability of a confined soil mass

A layer of sand 6 m thick lies beneath a clay stratum 5 m thick and above a bed of thick shale. In order to determine the permeability of the sand, a well was driven to the top of the shale and water pumped out at the rate of $10 \times 10^{-3}\,m^3/s$. Two observation wells were driven through the clay at 15 m and 30 m from the pump well and the water was found to rise to levels of 3 m and 2.5 m below the ground surface respectively. Calculate from first principles the coefficient of permeability of the soil (assuming Darcy's law). (HNC)

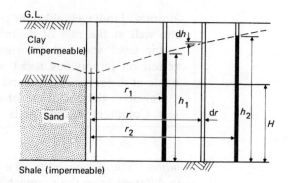

Figure 2.12

Solution   A section of the arrangement is shown on Fig. 2.12.

This is sometimes referred to as the artesian case since the water is confined under pressure between two impermeable layers.

Consider the inward flow through the boundary of a cylindrical section of radius r.

$$\text{Quantity flowing in unit time} = \frac{Q}{t} = q = \text{area} \times \text{velocity}$$

and assuming Darcy's law $v = k \cdot i$

$$q = 2\pi r H \times k \cdot \frac{dh}{dr} \qquad (H \text{ will be constant in this case})$$

$$\therefore \frac{dr}{r} = \frac{2\pi k H}{q} \cdot dh$$

Integrating between the appropriate limits

$$\int_{r_1}^{r_2} \frac{dr}{r} = \frac{2\pi k H}{q} \int_{h_1}^{h_2} dh$$

$$\therefore \log_e \frac{r_2}{r_1} = \frac{2\pi k H}{q}(h_2 - h_1)$$

$$\therefore k = \frac{q}{2\pi H(h_2 - h_1)} \cdot \log_e \frac{r_2}{r_1} \qquad (2.13)$$

From the given information

$$q = 10 \times 10^{-3} \text{ m}^3/\text{s}, \ H = 6 \text{ m}, \ h_2 = 11 - 2 \cdot 5 = 8 \cdot 5 \text{ m},$$
$$h_1 = 11 - 3 = 8 \text{ m}$$

Substituting in the above equation

$$k = \frac{10 \times 10^{-3}}{2\pi \times 6(8 \cdot 5 - 8 \cdot 0)} \log_e \frac{30}{15}$$

$$= 0.367 \times 10^{-3} \text{ m/s}$$

$$= \mathbf{0 \cdot 367 \ mm/s}$$

Flow nets

2.10 Drawing a flow net

Explain what is meant by a flow net. With the aid of a simple example describe how an approximate flow net can be drawn.

Also describe briefly an experimental method of deriving a flow net using the electrical analogy.　　　　(ICE)

Solution　The paths taken by moving particles of water as they flow through a permeable material may be represented pictorially by a series of *flow lines*. The flow lines are nearly parallel curved lines since water tends to take the shortest path from point to point but only to change direction in smooth curves [Fig. 2.13(*a*)].

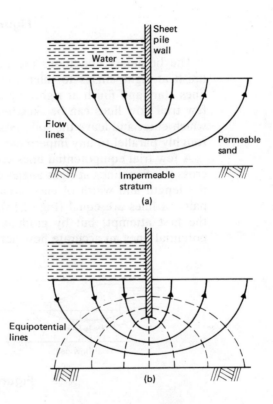

Figure 2.13

The points of equal head of water on each flow line can be joined to give another series of curves known as *equipotential lines* [Fig. 2.13(*b*)]. These lines cross the flow lines at right angles.

The pattern of approximate squares formed by these two sets of lines is known as a *flow net*.

To construct a flow net, a cross-section of the site and structure is drawn to scale [Fig. 2.14(*a*)].

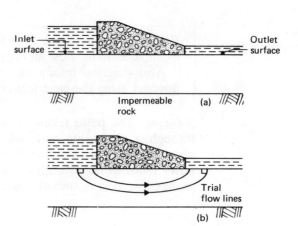

Figure 2.14

The boundary conditions are then examined. Flow lines start and finish at right angles to inlet and outlet surfaces; and equipotential lines start and finish at right angles to impermeable surfaces. Thus a few trial flow lines can be sketched in. These must be smooth curves which enter and leave the soil mass at right angles to its surface and be roughly parallel to any impermeable surfaces [Fig. 2.14(b)].

A few trial equipotential lines are then added to the diagram. These cross the flow lines at right angles and the aim is to draw them so that the length and width of each area enclosed by intersecting, adjacent pairs of lines are equal (Fig. 2.15). This is unlikely to be achieved at the first attempt, but by gradual adjustment of the flow and equipotential lines an accurate flow net can be drawn.

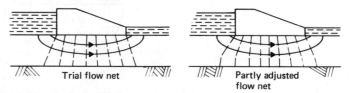

Figure 2.15

The flow of water through soil may be compared with the flow of electricity through a two-dimensional conducting medium. Thus an experimental method of deriving a flow net is to make use of a special type of electrosensitive conducting paper.

The section to be investigated is drawn to a convenient scale on the paper and the impermeable surfaces are cut out to prevent the flow of electricity across them. The boundary conditions are established by applying an electrical potential at the entry and exit of the flow. Then, using a Wheatstone bridge and a voltage probe, the points of equipotential can be traced and the equipotential lines plotted.

When the equipotential lines have been established, it is a straight-forward procedure to sketch in the corresponding flow lines and thus complete the flow net. The method lends itself to cases where the boundaries are somewhat irregular.

2.11 Flow net for a sheet pile wall

A sheet pile wall is driven to a depth of 6 m into permeable soil which extends to a depth of 13·5 m below ground level. Below this there is an impermeable stratum. There is a depth of water of 4·5 m on one side of the sheet pile wall. Make a neat sketch of the flow net and determine the approximate seepage under the sheet pile wall in m³/day, taking the permeability of the soil as 6×10^{-3} mm/s. (ICE)

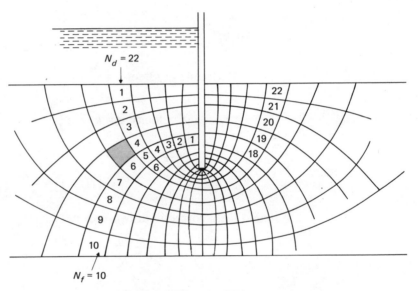

Figure 2.16

Solution The flow net is sketched by the method outlined in solution 2.10 and is shown on Fig. 2.16.

The flow net is used to determine the seepage under the sheet pile wall. If it has been drawn correctly, the drop in piezometric head (the height to which water would rise in an open stand-pipe at the point considered) will be constant between successive equipotential lines. (Fig. 2.17).

$$\therefore \quad \Delta h = \frac{h}{N_d}$$

where Δh = drop in head between any two adjacent equipotential lines

N_d = total number of head drops from inlet to outlet.

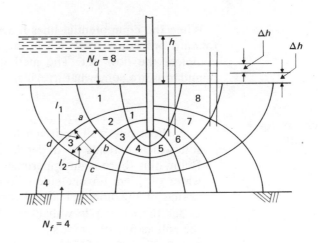

Figure 2.17

Consider a unit thickness of the portion of a flow channel *abcd*.

$$\text{Hydraulic gradient } i = \frac{\Delta h}{l_1} = \frac{h}{N_d \cdot l_1}$$

$$\text{Seepage velocity } v = k \cdot i = \frac{kh}{N_d \cdot l_1}$$

$$\text{Seepage } \Delta q = \text{area} \times \text{velocity}$$

$$= \left((l_2 \times 1) \times \frac{kh}{N_d \cdot l_1} \right)$$

and for a correctly drawn flow net $l_1 = l_2$

$$\therefore \ \Delta q = \frac{kh}{N_d}$$

$$\therefore \text{ Total seepage } \boldsymbol{q = kh\frac{N_f}{N_d}} \tag{2.14}$$

where N_f = number of flow squares across section.
From Fig. 2.16

$$N_f = 10 \quad N_d = 22$$

$$\therefore \ q = \frac{6 \times 10^{-3}}{10^3} \times 4 \cdot 5 \times \frac{10}{22} \, \text{m}^3/\text{s} \, \frac{60 \times 60 \times 24 \, \text{s}}{\text{day}} = \boldsymbol{1 \cdot 06 \, \text{m}^3/\text{day}}$$

2.12 Flow net for a dam

Figure 2.18 shows the cross-section of a concrete dam founded on slightly permeable soil, below which there is an impermeable stratum. Make a sketch of the flow net, assuming the soil to be isotropic and explain how this can be used to find the rate of seepage of water under the dam, if k for the soil $= 12.5 \times 10^{-3}$ mm/s.

What modifications would you make if the permeability in the horizontal direction was found to be four times that in the vertical direction? (ICE)

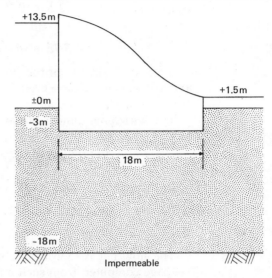

Figure 2.18

Solution The completed flow net is shown on Fig. 2.19. An isotropic soil is one which has the same properties in all directions.

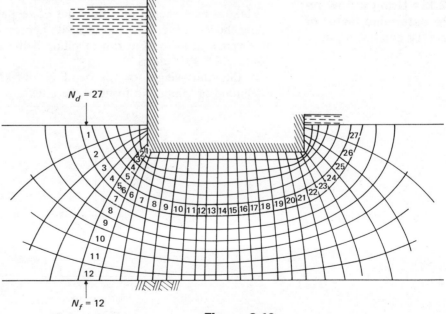

Figure 2.19

From the flow net

$$N_f = 12 \quad N_d = 27$$

$$\therefore \text{ seepage } q = kh\frac{N_f}{N_d}$$

$$= \frac{12 \cdot 5 \times 10^{-3}}{10^3} \times (13 \cdot 5 - 1 \cdot 5) \times \frac{12}{27} \frac{60 \times 60 \times 24 \text{ s}}{\text{day}}$$

$$= \textbf{5·75 m}^3\textbf{/day/m run of dam.}$$

If the permeability of the soil in the horizontal direction $= k_x$ and in the vertical direction $= k_z$, the sectional drawing of the site must be drawn to a distorted scale.

The horizontal dimensions are altered by the factor:

$$\sqrt{\frac{k_z}{k_x}} \quad \text{in relation to the vertical dimensions.}$$

In this case $k_x = 4kz$,

$$\therefore \text{ horizontal scale would be } = \sqrt{\tfrac{1}{4}} \text{ vertical scale}$$

$$= \tfrac{1}{2} \text{ vertical scale.}$$

The flow net is then drawn on this diagram in the usual way. The seepage computed from such a flow net will be the correct seepage for a soil having an average coefficient of permeability

$$k_{av} = \sqrt{k_x \cdot k_z} \text{ in both directions.}$$

2.13 Using a flow net to determine factor of safety against piping

Make a neat sketch of a flow net of seepage under the sheet piling shown in Fig. 2.20 and estimate approximately the quantity of seepage in m^3/min/m run of piling if the permeability of the sand is 18×10^{-3} mm/s.

If the unit weight of the sand is $18 \cdot 5$ kN/m^3 is there any likelihood of 'piping' in front of the piles?   (ICE)

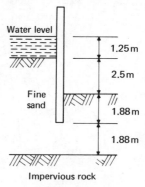

Water level

1.25 m

2.5 m

Fine sand

1.88 m

1.88 m

Impervious rock

Figure 2.20

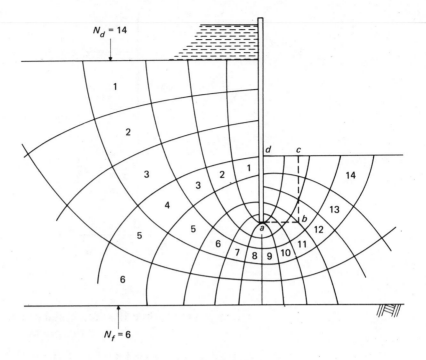

Figure 2.21

Solution The flow net is shown on Fig. 2.21.
From the figure:

$$N_f = 6 \quad N_d = 14$$

$$\therefore \ q = \frac{18 \times 10^{-3}}{10^{-3}} \times 3{\cdot}75 \times \frac{6}{14} \frac{\mathrm{m}^3}{\mathrm{s}} \left[\frac{60\ \mathrm{s}}{\mathrm{min}}\right]$$

$$= \mathbf{1{\cdot}730 \times 10^{-3}\ m^3/mm\ per\ m\ run\ of\ wall.}$$

Piping is the term applied to an unstable condition which can occur in a soil mass when the upward seepage pressure of the water is equal to the submerged weight of the soil above it. (See solution 2.3).

It has been found that in the case of a sheet pile wall, piping is most likely to occur within a width in front of the wall equal to about half the depth of penetration of the piles. It will be seen from the flow net that this is the place where the water is seeping upwards and the lines are very closely spaced—always a danger signal.

To determine the factor of safety against piping, consider the prism of soil of unit thickness *abcd* (Fig. 2.22).

$$ad = \text{depth of penetration of piles} = 1{\cdot}88\ \mathrm{m}$$

$$ab = \tfrac{1}{2}ad = 0{\cdot}94\ m$$

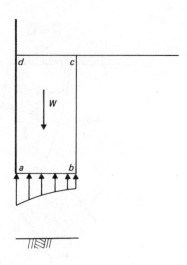

Figure 2.22

The downward force W = force due to submerged density of soil in the prism

$$= (\gamma_{sat} - \gamma_w) \times 1{\cdot}88 \times 0{\cdot}94 = (18{\cdot}5 - 10{\cdot}0) \times 1{\cdot}88 \times 0{\cdot}94$$
$$= 15 \text{ kN.}$$

The upward seepage force on plane ab is found from the flow net. The seepage pressure on ab will be equal to the loss of head between ab and cd in m of water. This head loss is equal to the average number of drops in head n_{av} between plane ab and plane cd divided by the total number of head drops from inlet to outlet N_d and multiplied by the total head loss h.

$$\therefore \text{ The upward seepage force} = \frac{n_{av} \cdot h}{N_d} \times \gamma_w \times \text{area of plane } ab$$

$$= \frac{4{\cdot}5 \times 3{\cdot}75}{14} \times 10{\cdot}0 \times (0{\cdot}94 \times 1)$$

$$= 11{\cdot}33 \text{ kN.}$$

$$\therefore \text{ Factor of safety against piping } F = \frac{\text{downward force}}{\text{upward force}}$$

$$= \frac{15}{11{\cdot}33} = \textbf{1·32.}$$

2.14 Flow net for a cofferdam

Figure 2.23 shows the cross-section of a long cofferdam into which the flow can be considered two-dimensional. Sketch the flow net (to the right of the centre-line only) for this situation.

The base of the soil stratum is at a considerable depth.

Determine the seepage into the cofferdam (per metre run), if the water level inside is maintained at excavated ground level. The coefficient of permeability of the soil is 0·015 m/s in every direction.

Using the flow net determine the distribution of water pressure (expressed as metres head of water) both on the outside and inside of the sheet piling and indicate the values on the left hand part of the drawing.

Comment on the stability of the proposed structure.   (ICE)

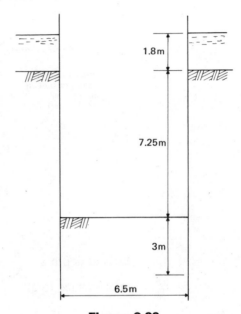

Figure 2.23

Solution   The flow net is shown on Fig. 2.24.

From the flow net: Number of head drops $N_d = 15$

Number of flow squares $N_f = 7$

$$\therefore \text{ Seepage } q = k \cdot h \cdot \frac{N_f}{N_d} \qquad (2.14)$$

$$= 0·015 \times 9·05 \times \tfrac{7}{15} = 0·063 \text{ m}^3/\text{s}.$$

This is for one side only and the total seepage from both sides

$$= 2 \times 0·063 = \mathbf{0·126 \, m^3/s.}$$

At any depth z below the ground surface the pore water pressure

= (head of water above z − loss of head as water flows through soil) γ_w

Total head loss = 9·05

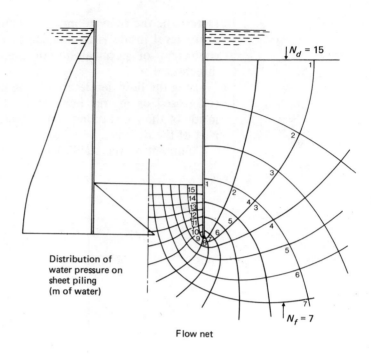

Distribution of
water pressure on
sheet piling
(m of water)

$N_d = 15$

$N_f = 7$

Flow net

Figure 2.24

$\therefore$ Loss at depth $z = \dfrac{n}{N_d} \times 9.05$ where n = no. of head drops from inlet surface to z.

$\therefore$ Pore water pressure $= \left(1.8 + z - \dfrac{n}{N_d} \times 9.05\right)\gamma_w$ kN/m^2

$\therefore$ Distribution of water pressure on sheet piling

$$= \left(1.8 + z - \dfrac{n}{N_d} \times 9.05\right)\gamma_w \text{ kN/m}^2 \text{ (since water pressure acts}$$
equally in all directions)

$$= \left(1.8 + z - \dfrac{n}{N_d} \times 9.05\right) \text{ m of water.}$$

Using this relationship, since $N_d = 15$

$$p = \left(1.8 + z - \dfrac{n}{15} \times 9.05\right) = (1.8 + z - 0.6n) \text{ m of water}$$

and by scaling the distances z corresponding to n off the flow net, the following results are obtained:

n	z (m)	$1 \cdot 8 + z - 0 \cdot 6n$ (m of water)	n	z (m)	$1 \cdot 8 + z - 0 \cdot 6n$ (m of water)	n	z (m)	$1 \cdot 8 + z - 0 \cdot 6n$ (m of water)
1	2·44	3·64	6	10·04	8·22	11	9·20	4·37
2	5·03	5·63	7	10·25	7·83	12	8·84	3·41
3	7·00	7·00	8	10·25	7·23	13	8·29	2·25
4	8·38	7·77	9	9·90	6·28	14	7·85	1·24
5	9·45	8·24	10	9·69	5·46	15	7·25	0

The values of water pressure are plotted on the left hand side of Fig. 2.24.

An approximate flow net is generally sufficient if quantities of seepage are required but if the pore water pressures are to be estimated from the flow net, it must be accurate.

From the flow net, the upward seepage pressure on the soil on the inside of the piling is approximately

$$\frac{6}{15} \times 9 \cdot 05 \times 10 = 36 \cdot 2 \text{ kN/m}^2. \text{ (See solution 2.11.)}$$

Taking the saturated unit weight of the soil $\gamma = 20$ kN/m^3

The downward pressure at the foot of the sheet piling due to the submerged weight of the soil $= (20 - 10) \times 3$
$$= 30 \text{ kN/m}^2$$

It would appear from this that there is a very distinct danger of piping occurring and the piles should be driven deeper.

Problems 2

1. The following information was obtained from a borehole:
  0 to −3 m, fine sand, saturated density 1·92 Mg/m^3
  −3 to −7·5 m, clay, saturated density   2·00 Mg/m^3
  below 7·5 m medium sand.

  The water table was at ground level. Calculate, and draw diagrams showing the variations of the following quantities with depth:
  (a) the total vertical pressure;
  (b) the effective vertical pressure.

  If the water table were later lowered by 2 m by pumping, show how the above quantities would be affected, assuming that the fine sand remained saturated with held water above the new water table. (HNC)

[(a) 57·6, 147·6 kN/m^2; (b) 27·6, 72·6 kN/m^2
σ unchanged; σ' increased by 20 kN/m^2]

2. A layer of saturated gravel ($\rho = 2\cdot00\,\text{Mg/m}^3$, $\rho_d = 1\cdot57\,\text{Mg/m}^3$) $2\cdot75$ m thick overlies 15 m of saturated clay ($\rho = 1\cdot84\,\text{Mg/m}^3$) which in turn rests on impermeable rock. The ground-water level, which is initially at the ground surface, is lowered rapidly by pumping, leaving the gravel dry.

Calculate the immediate changes in the pore water pressure and vertical effective stress at the base of the clay stratum due to this dewatering. (UL)

($u = -11\cdot8\,\text{kN/m}^2$; no change in σ' until consolidation occurs i.e. some considerable time later.)

3. (*a*) Show that water flowing through soil exerts, on unit volume of the soil, a force equal to the product of the hydraulic gradient and the unit weight of water.

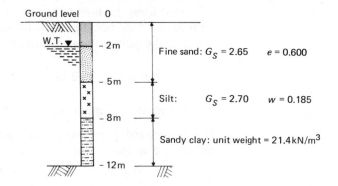

(*b*) A soil profile as shown in Fig. 2.25 was obtained during site investigation. Sketch graphs, showing the variations to a depth of 12 m of: (i) the vertical effective pressure, and (ii) the pore water pressure. Indicate salient values. (Assume that the sand above the water table is dry.) (SCOTEC)

(At -12 m (i) $143\cdot6\,\text{kN/m}^2$; (ii) $100\,\text{kN/m}^2$)

4. A permeameter of 82 mm diameter contains a sample of soil of length 350 mm. It can be used either for constant head or falling head tests. The standpipe used for the latter has a diameter of 25 mm. In a constant head test the loss of head was 1 150 mm measured on a length of 250 mm when the rate of flow was $2\cdot73 \times 10^3\,\text{mm}^3/\text{s}$. Find the coefficient of permeability of the soil in cm/s.

If a falling head test were made on the same soil, find what time would be taken for the head to fall from 1 500 mm to 1 000 mm.

Either work from first principles or derive the formulae used but assume Darcy's Law. (UL)

($11\cdot2 \times 10^{-3}\,\text{cm/s}$, $118\,\text{s}$)

5. A large open excavation was made into a stratum of clay, $\gamma_{sat} = 17 \cdot 6 \, kN/m^3$. When the depth of excavation reached $7 \cdot 5$ m, the bottom rose and was flooded with a mixture of sand and water. Subsequent boring showed that the clay was underlain by a bed of sand with its surface at a depth of $11 \cdot 25$ m. Find the height to which water would have risen from the sand into a borehole before excavation started.

($4 \cdot 65$ m below ground level.)

6. In a falling head permeameter the diameter and length of the sample are respectively 75 mm and 150 mm, the diameter of the standpipe being 12 mm. If the time taken for the head to fall from 600 mm to 300 mm is 193 s, find the coefficient of permeability of the sample in mm/s.

A well of $1 \cdot 25$ m diameter is sunk to a depth of $12 \cdot 5$ m in a bed of the same soil, the water table being $1 \cdot 5$ m below the surface. If the draw-down at a radius of $1 \cdot 6$ km is not to exceed $0 \cdot 6$ m and the depth of water in the well is not to fall below $1 \cdot 2$ m find the yield of the well in m^3/min, assuming a horizontal water table when no pumping is taking place.

Work from first principles throughout, assuming Darcy's Law, or prove any formula used. (UL)

($0 \cdot 0138$ mm/s; $4 \cdot 02 \times 10^{-3} \, m^3/min$)

7. (*a*) State the formula which expresses Darcy's law, explaining the symbols used.

(*b*) A pumping test was carried out at a level site, where 9 m of clay overlies a stratum of fine sand $1 \cdot 5$ m thick; immediately under the sand lies impermeable bedrock. After the pumping rate became steady at $12 \cdot 75$ l/s, the average water levels in the observation wells were respectively:

at 6 m radius, $4 \cdot 8$ m below ground level, and
at 15 m radius, $4 \cdot 2$ m below ground level.

Determine:

(i) the coefficient of permeability of the sand, proving any formula used;

(ii) the effective pressure, during pumping at the bottom of the sand stratum, at 6 m radius, if the total pressure at the bottom of the clay was $150 \, kN/m^2$.

Take the bulk density of the sand as $2 \cdot 05 \, Mg/m^3$. (SCOTEC)

($k = 2 \cdot 06 \times 10^{-3}$ m/s; $\sigma' = 124 \, kN/m^2$)

8. It is proposed to form a wall of considerable length by driving sheet piles to a depth of 10 m into a uniform horizontal sand deposit of depth 20 m. The sand rests on impermeable bedrock. The wall will impound on one side a depth of 5 m of water above the level surface of the sand. On the other side an impervious concrete apron extending 20 m from the wall, will be laid on top of the sand. The down-stream water level is maintained at the ground surface.

Draw a flow net for this situation. Plot a curve showing the excess hydrostatic uplift pressures on the apron, and estimate the rate of seepage. Assume that the wall is impermeable and that the sand has a permeability of 0·006 m/s.

Comment on the stability of the proposals.   (ICE)

$(20 \times 10^{-3} \, \text{m}^3/\text{s approx.})$

9. What do you understand by the term 'flow net'?

A 12-m thick deposit of cohesionless soil of permeability $3\cdot5 \times 10^{-6}$ m/s has a level surface and overlies an impermeable layer. A long row of sheet piles is driven 6 m into the soil. The wall extends above the surface of the soil, and impounds a depth of 3·6 m of water on one side; the water level on the other side is maintained at ground level. Sketch the flow net and determine the seepage quantity per metre run of wall, deriving any formula used.

What is the value of the pore water pressure at a point near the toe of the wall?

How would you investigate the factor of safety against piping in this problem?   (ICE)

$(\text{Seepage} = 0\cdot33 \times 10^{-6} \, \text{m}^3/\text{s} \quad u = 77\cdot5 \, \text{kN/m}^2$

$\qquad\qquad = 0\cdot284 \, \text{m}^3/\text{day})$

3

Compressibility of soil and settlement of foundations

Stress distribution from external loading

When an area of soil is loaded, the vertical stresses within the soil mass will be increased. The increase is greatest directly under the loaded area but tends to extend indefinitely in all directions.

In order to estimate these stresses, it is assumed that the soil is elastic, homogeneous and isotropic. Then, using the theory of elasticity, the following formula has been developed by Boussinesq:

$$\sigma_z = \frac{3P}{2\pi z^2} \cdot \frac{1}{[1 + (r/z)^2]^{5/2}} \tag{3.1}$$

where σ_z = vertical normal compressive stress
 P = a vertical concentrated load
 z = vertical distance below P
 r = horizontal distance from P

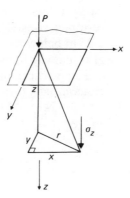

Figure 3.1

A natural soil is unlikely to comply with any of these assumptions and this should be borne in mind when using the formula.

It can be expressed in the form:

$$\sigma_z = \frac{KP}{z^2}$$

where K is an *influence factor* $= \dfrac{3}{2\pi} \dfrac{1}{[1 + (r/z)^2]^{5/2}}$.

Table 3.1 gives the values of K for various values of r/z.

Table 3.1 Boussinesq's influence factor for a point load

r/z	K	r/z	K	r/z	K
0·00	0·4775	1·00	0·0844	2·00	0·0085
0·10	0·4657	1·10	0·0658	2·10	0·0070
0·20	0·4329	1·20	0·0513	2·20	0·0058
0·30	0·3849	1·30	0·0402	2·30	0·0048
0·40	0·3294	1·40	0·0317	2·40	0·0040
0·50	0·2733	1·50	0·0251	2·50	0·0034
0·60	0·2214	1·60	0·0200	2·60	0·0029
0·70	0·1762	1·70	0·0160	2·70	0·0024
0·80	0·1386	1·80	0·0129	2·80	0·0021
0·90	0·1083	1·90	0·0105	2·90	0·0018

Worked examples

3.1 Stress in a soil mass due to point loads

Three point loads of 640, 160 and 320 kN, 2 m apart in a straight line, act at the surface of a soil mass. Calculate the resultant stresses produced by these loads on a horizontal plane 1·25 m below the surface, at points vertically below the loads and also halfway between them.

The vertical pressure σ_z due to a point load P is given by Boussinesq's formula:

$$\sigma_z = \frac{KP}{z^2}$$

The values of K are as follows:

r/z	0	0·8	1·6	2·4	3·2
K	0·4775	0·1386	0·0200	0·0040	0·0012

Sketch the curve showing the distribution of these resultant stresses at that level. (UL)

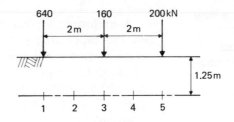

Figure 3.2

Solution   To find the vertical stresses due to any load, find the value of r/z and hence the corresponding value of K. Then σ_z is found from the formula given.

| | P | 640 kN *load* | | | 160 kN *load* | | | 320 kN *load* | | |
| z | | | | | | | | | | |
Point	r/z	K	σ_z (kN/m²)	r/z	K	σ_z (kN/m²)	r/z	K	σ_z (kN/m²)
1	0	0·4775	195·6	1·6	0·0200	2·1	3·2	0·0012	0·2
2	0·8	0·1386	56·8	0·8	0·1386	14·2	2·4	0·0040	0·8
3	1·6	0·0200	8·2	0	0·4775	48·9	1·6	0·0200	4·1
4	2·4	0·0040	1·6	0·8	0·1386	14·2	0·8	0·1386	28·4
5	3·2	0·0012	0·5	1·6	0·0200	2·1	0	0·4775	97·8

The total stress at any point is found by superposition of the individual stresses.

$$\text{kN/m}^2$$

$$\therefore \text{ Total stress at point } 1 = 195\!\cdot\!6 + 2\!\cdot\!1 + 0\!\cdot\!2 = 197\!\cdot\!9$$
$$2 = 58\!\cdot\!8 + 14\!\cdot\!2 + 0\!\cdot\!8 = 73\!\cdot\!8$$
$$3 = 8\!\cdot\!2 + 48\!\cdot\!9 + 4\!\cdot\!1 = 61\!\cdot\!2$$
$$4 = 1\!\cdot\!6 + 14\!\cdot\!2 + 28\!\cdot\!4 = 44\!\cdot\!2$$
$$5 = 0\!\cdot\!5 + 2\!\cdot\!1 + 97\!\cdot\!8 = 100\!\cdot\!4$$

The curve showing the distribution of these stresses is shown on Fig. 3.3.

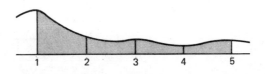

Figure 3.3

3.2 Stress in a soil mass due to distributed load using a Fadum chart

A uniformly loaded flexible footing imposes uniform loading on the supporting soil. Explain why the uniformity no longer applies when a soil supports a stiff footing, indicating the nature of the distribution of the contact pressure when the soil is (a) a clay (b) a sand.

A stiff footing of size 4.5×4.5 m carries a total uniformly distributed load of 3 250 kN. Using the chart on Fig. 3.6 find the intensity of vertical stress due to the loading at a point located at 4·5 m below a corner of the footing. (The contact pressure distribution may be approximated by assuming that a central area 3 m $\times 3$ m is loaded at twice the intensity of the remainder). Compare this value with that found assuming the whole load to be concentrated at the centre of the footing.   (ICE)

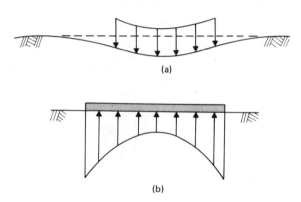

(a)

(b)

Figure 3.4

Solution   (a) When a cohesive soil is subjected to a uniform pressure the loaded area and adjacent areas will deform as shown (to an exaggerated scale) in Fig. 3.4(a).

With a rigid foundation, the soil tends to deform in the same way, but since the foundation does not bend correspondingly, a redistribution of pressure under the loaded area takes place as shown in Fig. 3.4(b).

(b) In the case of cohesionless soil the shear strength and modulus of elasticity tend to increase with increases in the confining pressure and thus a uniform pressure will cause the soil to deform as shown in Fig. 3.5(a) since the soil at the edges is less confined than that at the centre.

(a)   (b)

Figure 3.5

Thus under a rigid foundation the stress redistribution is as shown on Fig. 3.5(*b*).

In a development of Boussinesq's theory for a point load, the vertical stress at any depth z under the corner of a rectangle of dimensions $B \times L$ and carrying a uniform load q/unit area can be expressed in the form:

$$\sigma_z = qI$$

where I is the influence factor

$$\frac{1}{4}\left[\frac{2mn(m^2 + n^2 + 1)^{1/2}}{m^2 + n^2 + m^2n^2 + 1} - \frac{m^2 + n^2 + 2}{m^2 + n^2 + 1} - \tan^{-1}\frac{2mn(m^2 + n^2 + 1)^{1/2}}{m^2 + n^2 + 1 - m^2n^2} \right]$$

and $m = \dfrac{B}{z}$ $n = \dfrac{L}{z}$.

The values of I have been calculated and are shown on Fig. 3.6 known as *Fadum's chart*.

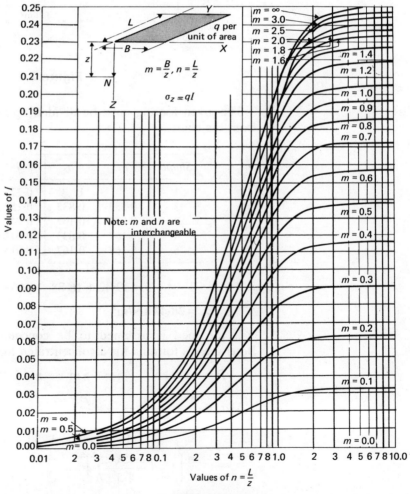

Figure 3.6

For the given footing, the total load is 3 250 kN. If the pressure distribution under the central area is $2x$ kN/m^2, the pressure under the outer rectangle $= x$ kN/m^2

$$\therefore \ 3^2 \times 2x + (4 \cdot 5^2 - 3^2)x = 3\,250$$

$$\therefore \ x = 111 \text{ kN/m}^2$$

In order to use the chart, the figure must be split up into a number of rectangles or squares which have one corner over the point where the stress is required.

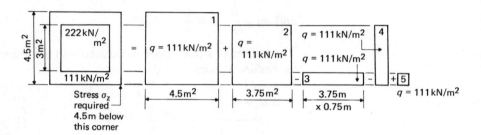

Figure 3.7

The given footing must be divided up as shown on Fig. 3.7. Then from the chart:

Shape	m	n	I	$\sigma_z = qI$ kN/m^2
1	1	1	0·175	19·42
2	0·833	0·833	0·152	16·88
3	0·833	0·167	0·042	4·66
4	0·167	0·833	0·042	4·66
5	0·167	0·167	0·013	1·44

$$\therefore \ 1 + 2 - 3 - 4 + 5 = 19\cdot42 + 16\cdot88 - 4\cdot66 - 4\cdot66 + 1\cdot44$$

$$\therefore \ \sigma_z = \mathbf{28\cdot42 \ kN/m^2}$$

If the load is assumed to be concentrated at the centre, using Boussinesq's formula for a concentrated load:

$$P = 3\,250 \text{ kN} \qquad z = 4 \cdot 5 \text{ m} \qquad r = 2 \cdot 25\sqrt{2} \ m$$

$$\sigma_z = \frac{3}{2\pi} \frac{1}{[1 + (r/z)^2]^{5/2}} \times \frac{P}{z^2}$$

$$= \frac{3}{2\pi} \frac{1}{\left[1 + \left(\dfrac{2\cdot25\sqrt{2}}{4\cdot5}\right)^2\right]^{5/2}} \times \frac{3\,250}{4\cdot5^2}$$

$$= \mathbf{27\cdot7 \ kN/m^2.}$$

3.3 Relief of stress in a soil mass due to excavation

A pit 10 m square and 7·5 m deep is to be excavated in a soil of unit weight 20 kN/m³. Plot a profile of the relief of vertical stress produced below the centre point of the excavation to a depth of 10 m below the base of the excavation.

Solution The effect of excavating the soil is the reverse of an applied uniform load. The stress relief at various depths can thus be calculated using the influence factors I obtained from Fig. 3.10.

Relief of pressure at base of excavation $q = \gamma \times z$

$$= 20 \times 7 \cdot 5 = 150 \text{ kN/m}^2$$

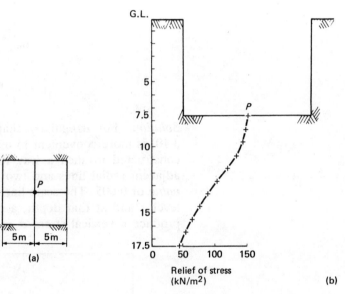

Figure 3.8

The pit can be divided up into four squares $5 \text{ m} \times 5 \text{ m}$, each with a corner at the centre of the excavation (Fig. 3.8(a)). Then for each shape:

Depth z (m)	m	n	I	$\sigma_z = 4 \times qI$ (kN/m²)
1	5·00	5·00	0·246	147·6
2	2·50	2·50	0·240	144·0
3	1·67	1·67	0·225	135·0
4	1·25	1·25	0·202	121·0
5	1·00	1·00	0·176	105·6
6	0·83	0·83	0·150	90·0
7	0·71	0·71	0·130	78·0
8	0·62	0·62	0·110	66·0
9	0·55	0·55	0·093	55·8
10	0·50	0·50	0·082	49·2

The resulting relief of pressure is plotted on Fig. 3.8(*b*).

3.4 Stress at a point in a soil mass using a Newmark chart

A pipe at a depth of 4 m passes close to a foundation as shown on Fig. 3.9. The foundation exerts a pressure of 40 kN/m² on the ground. Using the Newmark Chart shown in Fig. 3.10, determine the pressure exerted on the pipe by the foundation at point X. Check the result using the Fadum Chart (Fig. 3.7).

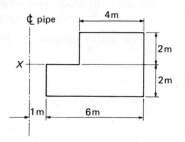

Figure 3.9

Solution   For irregularly shaped figures, the Newmark Chart (Fig. 3.10) is more convenient to use than the Fadum Chart (Fig. 3.6). It is constructed in such a way that each sub-division, bounded by two adjacent radial lines and two adjacent circles, represents an *influence value* of 0·005. The scale line AB is equal to the depth below ground level z and at that depth, a pressure of q kN/m² on the surface will produce a vertical stress $\sigma_z = 0·005q$ kN/m² at point N. Thus to find

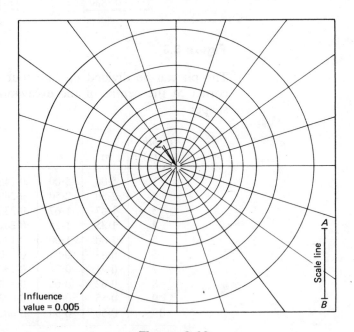

Figure 3.10

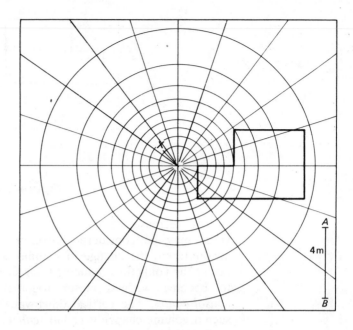

Figure 3.11

the stress at a depth z beneath any point X on a foundation, the plan area is drawn to a scale such that $AB = z$ and the point X is superimposed on the chart. Then by counting up the number of influence areas (including fractional ones) covered by the foundation, the stress can be calculated.

In the given example, AB on the scale line represents 4 m and the plan of the foundation is drawn to this scale and point X on the pipe placed at the centre of the Newmark Chart (Fig. 3.11).

The number of influence areas covered by the foundation = 21·5

$$\therefore \text{ Vertical stress } \sigma_z = 0.005 \times 21.5 \times 40 = \mathbf{4.3 \, kN/m^2}$$

Using the Fadum Chart, the foundation must be split up into the shapes shown on Fig. 3.12. Then from the chart:

Shape	m	n	I	$\sigma_z = qI$ (kN/m²)
1	0·5	1·75	0·133	5·32
2	0·5	1·75	0·133	5·32
3	0·5	0·25	0·048	1·92
4	0·5	0·75	0·109	4·36

$$\therefore \text{ Vertical stress} = 1 + 2 - 3 - 4 = 10.64 - 6.28 = \mathbf{4.36 \, kN/m^2}$$

Consolidation   When a saturated soil is subjected to a steady pressure e.g. due to the weight of overlying soil or to the load from a foundation, its volume

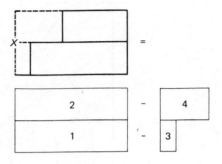

Figure 3.12

will be decreased. Since both the soil particles and the water in the voids may be considered incompressible at the sort of pressures encountered, the change in volume can only occur if water is forced out of the voids thus reducing their size and enabling the solid particles to become wedged closer together. This process is known as *consolidation.* The vertical downward displacement brought about by such a volume change is called *settlement.*

The amount of consolidation for a unit increase in pressure depends on a property of the soil known as *compressibility.* The rate at which the consolidation takes place depends on the ease with which water can escape from the soil and is therefore related to the *permeability* of the soil. These two effects are combined in a composite factor for the soil known as the *coefficient of consolidation.*

3.5 Estimating pre-compression stress in a soil mass

Explain what is meant by:
(*a*) a normally consolidated clay stratum,
(*b*) an overconsolidated clay stratum.
Sketch curves showing the variation of void ratio with pressure increase in each case and show how the pre-compression stress can be estimated.   (HNC)

Solution (*a*) A clay stratum is formed by the gradual deposition of very fine particles under water. Under normal conditions, therefore, it will be fully saturated and the gradual increase in pressure due to the weight of material deposited above it will cause it to be steadily consolidated. Under these conditions, the stratum is said to be *normally consolidated.*

Figure 3.13(*a*) shows the variation in void ratio *e* with increase of pressure *p*.

(*b*) It is possible in the course of the geological history of a clay stratum for some of the overburden pressure, which has caused the stratum to consolidate, to be removed e.g. by soil erosion. In these circumstances the soil may have been subjected to a pressure greatly in excess of its present overburden and is said to be *overconsolidated.*

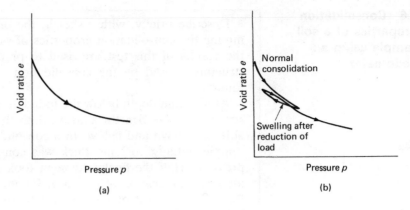

Figure 3.13

The variation in void ratio e with pressure increase p for an overconsolidated clay is shown on Fig. 3.13(*b*).

It is of some importance when considering the possible settlement of a stratum under load, to know how much pressure an overconsolidated layer has been subjected to in the past, since under the same loading a normally consolidated soil will tend to compress more than an overconsolidated soil.

The amount of pre-compression may be estimated by means of an empirical construction devised by Casagrande.

The e/p curve is plotted to a semi-logarithmic scale and a typical result is shown on Fig. 3.14. The construction is as follows.

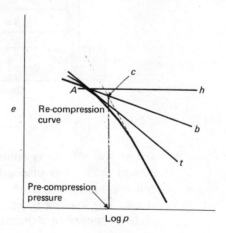

Figure 3.14

Locate the point of greatest curvature on the re-compression branch by eye (point A). Through this point draw a tangent At and a horizontal line Ah. Then bisect angle hAt by line Ab. Produce the straight line portion of the curve back to meet Ab at c. The pressure at this point is taken as the pre-compression stress on the layer.

3.6 Consolidation
properties of a soil
sample using an
oedometer

Describe briefly, with a sketch, the oedometer test for determining the consolidation properties of saturated clay. Show how
the results of this test are used to predict the settlement of a
structure caused by the consolidation of clay soil beneath the
foundation.

At a certain depth below the foundation of a building there is a
stratum of clay 6 m thick with relatively incompressible permeable soil above and below. In a consolidation test on this clay a
sample initially 26·2 mm thick was compressed under a steady
pressure. Half the final settlement took place in the first 11 min
after the application of pressure. Estimate how long it will take
for the settlement of the building to reach 50% of its ultimate
value.   (ICE)

Solution   Figure 3.15 shows the consolidation test cell known as an
oedometer.

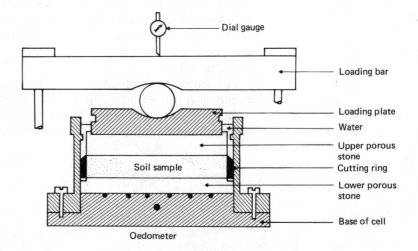

Figure 3.15

The soil sample is contained in a cutting ring 760 mm diameter by
20 mm thick. It is placed between two porous stones inside a circular
cell containing water. A loading plate is placed on the upper porous
stone. The loading beam is then brought into contact with the loading
plate by means of a thumb screw and a dial gauge set at zero is placed
on the beam.

The loading beam is counterbalanced and connected to the weight
hangers in such a way that the load can be applied instantaneously to
the sample.

When the first load is applied, readings of the dial gauge are taken
at $\frac{1}{4}, \frac{1}{2}$, 1, 2, 4 min; $\frac{1}{4}, \frac{1}{2}$, 1, 2, 4 and 24 h from the start. After 24 h, the
load on the specimen is increased and a new set of dial gauge readings

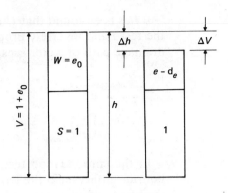

Figure 3.16

is obtained. This is repeated for a number of load increments. Finally the load is removed and the sample allowed to swell and the amount of recovery is measured. The sample is then removed from the cell and its moisture content found.

Consider the soil-phase diagram shown in Fig. 3.16. If it is subjected to a steady pressure the volume of the voids will decrease due to the expulsion of water but the volume of the solid particles will not change.

$$\therefore \text{ Unit volume change } \frac{\Delta V}{V} = \frac{de}{1 + e_0}$$

If the cross-sectional area remains unchanged

$$\frac{\Delta h}{h} = \frac{\Delta V}{V} = \frac{de}{1 + e_0}$$

This is the case in a consolidation test. Thus if the sample is subjected to a pressure p and the change in the void ratio is de then:

$$\text{Change in thickness of sample } dh = h \times \frac{de}{1 + e_0} \tag{3.2}$$

In the case of a stratum of the same soil of thickness z subjected to the same pressure p, the change in the void ratio of the soil will be the same and then:

$$\text{Change in thickness of stratum} = z \times \frac{de}{1 + e_0}$$

$$= \text{settlement } \rho_v$$

The appropriate values of de and $1 + e_0$ are found from the results of an oedometer test.

The *degree of consolidation* of a soil

$$= U = \frac{\text{consolidation after time } t}{\text{total consolidation}}$$

It has been found that U is proportional to the time from the start of the test t and inversely proportional to the square of the length of the drainage path of water escaping from the soil d.

$$\therefore\ U \propto \frac{t}{d^2}$$

$$\therefore\ U \propto \left(\frac{t}{d^2}\right)_{\text{sample}} \propto \left(\frac{t}{d^2}\right)_{\text{soil}}$$

In the consolidation test the soil is free to drain from the top and bottom surfaces and the length of the drainage path is taken as half the thickness of the sample. In the given conditions, the soil has permeable layers above and below and the length of the drainage path will therefore be half the thickness of the stratum.

$$\therefore\ \text{for } U = 50\%$$

For sample $t = 11$ min $d = \dfrac{26 \cdot 2}{2} = 13 \cdot 1$ mm

For soil $t = ?$ $d = \dfrac{6 \times 10^3}{2} = 3 \times 10^3$ mm

$$\therefore\ \left(\frac{11}{13 \cdot 1^2}\right)_{\text{sample}} = \left(\frac{t}{(3 \times 10^3)^2}\right)_{\text{soil}}$$

$$\therefore\ t_{\text{soil}} = \frac{11 \times (3 \times 10^3)^2}{(13 \cdot 1)^2} \text{ min} \left[\frac{\text{day}}{60 \times 24 \text{ min}}\right]$$

$$= \textbf{400 days}$$

3.7 Coefficient of compressibility of a soil sample

The following results were recorded from a consolidation test on a sample of saturated clay, each pressure being maintained constant for 24 h.

Pressure (kN/m²)	Thickness of sample after 24 h (mm)
0	25·0
50	24·6
100	24·4
200	24·2
400	23·9
800	23·7
0	24·2

The water content at the end of the test was 23·1% and the specific gravity of the soil particles was 2·68. Calculate the void ratio at the end of each pressure stage, and plot the pressure/void ratio curve.

Find from the curve the coefficient of compressibility when the pressure is $400\,\text{kN/m}^2$. (ICE)

Solution In solution 3.5 it was shown that in a consolidation test, since the cross-sectional area of the sample does not change, the change in thickness of a sample

$$dh = h \times \frac{de}{1 + e_0} \qquad (3.2)$$

$$\therefore\ de = \frac{1 + e_0}{h}.\, dh$$

$\therefore$ The change in the void ratio at the end of each pressure stage can be calculated from the change in thickness of the sample during the corresponding pressure stage.

For a *saturated* soil, the water content

$$w = \frac{M_W}{M_S}.\, 100$$

$$\frac{M_W}{M_S} = \frac{V_W \cdot \rho_w}{V_S \cdot \rho_w \cdot G_s} \quad \text{but} \quad \frac{V_W}{V_S} = e$$

$$\therefore\ e = \frac{w \cdot G_s}{100} \qquad (3.3)$$

At the end of the test $w = 23.1\%$, $G_s = 2.68$

$$\therefore\ e = \frac{23.1 \times 2.68}{100} = 0.62 \quad \text{at } h = 24.2\,\text{mm}$$

$$\therefore\ de = \frac{1 + 0.62}{24.2}.\, dh = \mathbf{0.067\ dh}$$

Using this relationship the value of e at each pressure stage can be calculated. The values are tabulated below:

p (kN/m^2)	0·0	50	100	200	400	800	0
dh (mm)	0·8	0·4	0·2	0·0	−0·3	−0·5	0
de	0·054	0·027	0·013	0·0	−0·020	−0·034	0·0
$e = 0.62 + de$	0·674	0·647	0·633	0·62	0·600	0·586	0·620

These values are plotted on Fig. 3.17.

The *compressibility* of a soil $a_v = \dfrac{-de}{dp}$ $\qquad (3.4)$

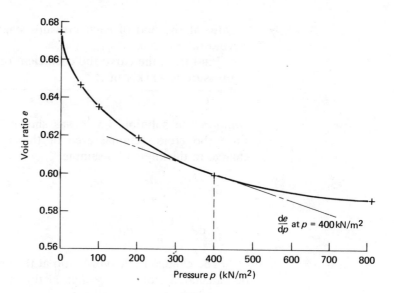

Figure 3.17

The *coefficient of volume compressibility* (or *volume decrease*) m_v is defined as the decrease in unit volume per unit increase in pressure.

$$\therefore \ m_v = -\frac{1}{h}\frac{\mathrm{d}h}{\mathrm{d}p} = -\frac{\mathrm{d}e}{(1+e_0)\,\mathrm{d}p} \tag{3.5}$$

$\dfrac{\mathrm{d}e}{\mathrm{d}p}$ represents the slope of the e/p curve at any selected point.

From Fig. 3.17 at $p = 400\ \text{kN/m}^2$ $\dfrac{\mathrm{d}e}{\mathrm{d}p} = -64 \times 10^{-6}$

$$\therefore \ m_v = \frac{1}{(1+0{\cdot}600)} \times 64 \times 10^{-6} = \mathbf{40 \times 10^{-6}\ m^2/kN}$$

It may be noted that it is not necessary to determine the values of e from a consolidation test, in order to find the value of m_v.

$$m_v = \frac{-\mathrm{d}e}{(1+e_0)\,\mathrm{d}p} \quad \text{and} \quad \mathrm{d}h = \frac{\mathrm{d}e}{(1+e_0)} \times h$$

$$\therefore \ m_v = -\frac{1}{h}\cdot\frac{\mathrm{d}h}{\mathrm{d}p} \tag{3.6}$$

Thus if the thickness h at each stage is plotted against the corresponding value of p, the slope of the curve at any point $= \dfrac{\mathrm{d}h}{\mathrm{d}p}$ and the value of m_v can be found (Fig. 3.18).

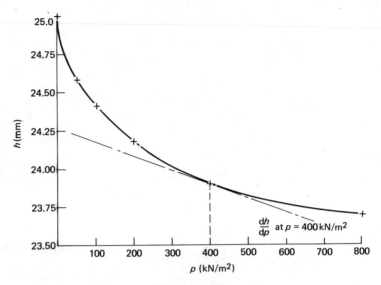

Figure 3.18

From this curve $\dfrac{\mathrm{d}h}{\mathrm{d}p} = -930 \times 10^{-6}$ at $p = 400 \text{ kN/m}^2$

$$\therefore m_v = \frac{930 \times 10^{-6}}{23 \cdot 9} = \mathbf{39 \times 10^{-6} \, m^2/kN}$$

3.8 Settlement of a foundation using results of oedometer test

A borehole at the site of a proposed building indicates the following strata (levels referred to ground level as datum);

Datum to -5 m	Compact sand $\rho = 2 \cdot 00 \text{ Mg/m}^3$, Ground-water level at $-3 \cdot 5$ m
-5 m to -10 m	Clay $\rho = 1 \cdot 92 \text{ Mg/m}^3$
Below -10 m	Impervious shale

The base of the foundation for the building is to be at -3 m and the additional pressure on the clay due to the weight of the building is estimated to be 100 kN/m^2 at the top of the clay and 40 kN/m^2 at the bottom of the clay. Calculate the probable final settlement of the clay, assuming the following results from oedometer tests.

Pressure (kN/m^2)	Void ratio
50	0·945
100	0·895
200	0·815
400	0·750

(ICE)

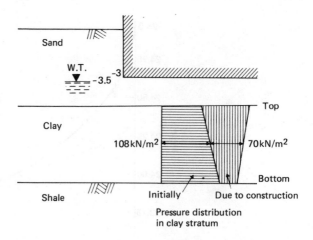

Sand

W.T.
▼ -3.5 -3

Top

Clay

108 kN/m²   70 kN/m²

Bottom

Shale   Initially   Due to construction

Pressure distribution
in clay stratum

Figure 3.19

Solution A sketch of the site is shown on Fig. 3.19. It should be noted that when a load is initially applied to a saturated soil, it has the effect of increasing the pore water pressure. As the process of consolidation continues and water is forced out of the pores, the pore water pressure drops and the effective pressure on the soil increases. Thus in consolidation problems, the pressures which have to be considered are the effective pressures.

Before construction:

Top of clay	kN/m²
Total vertical pressure = $(2 \cdot 00 \times 10) \times 5 =$	100
Pore water pressure = $(1 \cdot 00 \times 10) \times 1 \cdot 5 =$	15
∴ Effective pressure	85

Bottom of clay	kN/m²
Total vertical pressure = $100 + (1 \cdot 92 \times 10) \times 5 = 196$	
Pore Water pressure = $(1 \cdot 00 \times 10) \times 6 \cdot 5 =$	65
∴ Effective pressure	131

Assuming a linear distribution of pressure across the stratum, the average pressure on the clay before construction

$$= \frac{85 + 131}{2} = 108 \text{ kN/m}^2$$

After construction the average increase in pressure on the clay

$$= \frac{100 + 40}{2} = 70 \text{ kN/m}^2$$

and the final average effective pressure on the clay $= 178 \text{ kN/m}^2$. The graph of e/p is shown on Fig. 3.20.

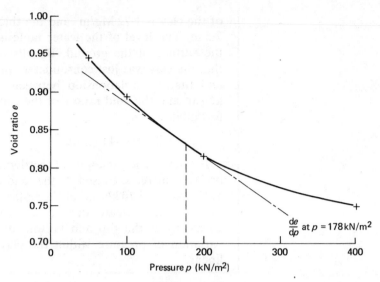

Figure 3.20

From this graph $e = 0.883$ at $p = 108 \text{ kN/m}^2$

$$\frac{de}{dp} = 716 \times 10^{-6}$$

$$\therefore \ m_v = \frac{1}{(1 + 0.883)} \cdot 716 \times 10^{-6} = \mathbf{380 \times 10^{-6} \ m^2/kN.}$$

$\therefore$ Settlement $\rho_v = m_v \times$ average increase in pressure $\times$ thickness of stratum

$$= 380 \times 10^{-6} \times 70 \times (5 \times 10^3)$$

$$= \mathbf{133 \ mm.}$$

m_v may also be calculated approximately by assuming a straight line relationship between e and p over the pressure increment

Then $\quad m_v = \dfrac{-e_1 - e_2}{p_1 - p_2} \cdot \dfrac{1}{1 + e_{\text{av}}}.$

In this example

$$m_v = -\frac{0.883 - 0.833}{108 - 178} \times \frac{1}{1 + 0.858} = \mathbf{384 \times 10^{-6} \ m^2/kN}$$

3.9 Settlement of a foundation using compression index of soil

A foundation is constructed at a depth of 4·5 m below the surface of a bed of compact incompressible sand of density 2·00 Mg/m³, which extends to a depth of 4·5 m below the base of the foundation. Under the sand there is a stratum of compressible clay which in turn rests on incompressible sand. The density of the clay is $1.72\,\text{Mg/m}^3$ and the thickness of the clay stratum is 7.5 m. The level of the water table is at a depth of 7.5 m below the surface of the ground. Results of oedometer tests revealed that the clay was just consolidated under the original overburden and that the relationship between the effective pressure p in kN/m^2 and the void ratio e of the clay could be expressed by the formula:

$$e = 1.2 - 0.044 \log_{10} p.$$

Estimate the amount of settlement expected if the gross pressure increase caused by the load of the structure is $75\,\text{kN/m}^2$ at the top and $10\,\text{kN/m}^2$ at the bottom of the clay stratum, and if the pressure release due to the excavation is $35\,\text{kN/m}^2$ and $10\,\text{kN/m}^2$ at the top and bottom of the clay respectively. The variation of pressure within the clay stratum is assumed to be linear.   (UL)

Solution   Using the results obtained from an oedometer test, if the void ratio e is plotted against the log of the corresponding value of p, the resulting curve will be of the form shown on Fig. 3.21 for a normally consolidated clay.

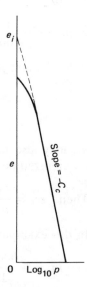

Figure 3.21

It can be seen that it is straight for an appreciable length. The straight length can be expressed by the formula

$$e = e_i - C_c \cdot \log_{10} p$$

where C_c is a dimensionless number known as the *compression index* and is the gradient of the straight portion; e_i is the void ratio at unit pressure.

This is the basis of the equation given in the question. Since C_c is constant for a wide range of pressure changes on any particular soil, it provides a convenient way of calculating the change in void ratio and hence the settlement due to a change in effective pressure.

For the given conditions:

Initially

At top of the clay: kN/m^2

Total overburden pressure	$\sigma = (2 \times 10) \times 9 =$	180
Pore water pressure	$u = (1 \times 10) \times 1 \cdot 5 =$	15
$\therefore$ Effective pressure	$\sigma' = \sigma - u =$	165

At bottom of clay kN/m^2

Total overburden pressure $= 180 + (1 \cdot 72 \times 10) \times 7 \cdot 5 = 309$
Pore water pressure $= (1 \cdot 0 \times 10) \times 9 =$ 90
$\therefore$ Effective pressure $=$ 219

$\therefore$ Average effective pressure at centre of clay stratum

$$= \frac{165 + 219}{2} = 192 \text{ kN/m}^2$$

$\therefore e_0 = 1 \cdot 2 - 0 \cdot 044 \log_{10} 192$

$\qquad = 1 \cdot 100.$

After construction of foundation:

At top of clay kN/m^2

Initial effective pressure $=$	165·0
Pressure release due to excavation $=$	35·0
$\therefore$ Net effective pressure $=$	130·0
Increase in pressure from structure $=$	75·0
$\therefore$ Final effective pressure $=$	205·0

At bottom of clay kN/m^2

Initial effective pressure $=$	219·0
Pressure release due to excavation $=$	10·0
	209·0
Increase in pressure from structure $=$	10·0
$\therefore$ Final effective pressure $=$	219·0

$\therefore$ Average effective pressure at centre of clay stratum

$$= \frac{205 + 219}{2} = 212$$

$\therefore e_f = 1 \cdot 2 - 0 \cdot 044 \log_{10} 212 = 1 \cdot 098$

$$\therefore \text{ Settlement } \rho_v = z \times \frac{de}{1 + e_0}$$

$$= (7 \cdot 5 \times 10^3) \cdot \frac{(1 \cdot 100 - 1 \cdot 098)}{1 + 1 \cdot 100} = \textbf{7·14 mm}$$

Rate of consolidation

When a pressure increase (p) is applied to a soil sample in an oedometer, the pressure will initially be carried entirely by the pore water [Fig. 3.22(b)]. This increase in the pore water pressure (u) causes the water to drain out of the sample both upwards and downwards. As the water escapes some of the pressure will be transferred to the soil grains, leading to an increase in the effective pressure (p'), [Fig. 3.22(c)] and a reduction in the thickness of the sample. The process of consolidation will be complete when $u = 0$ and $p' = p$.

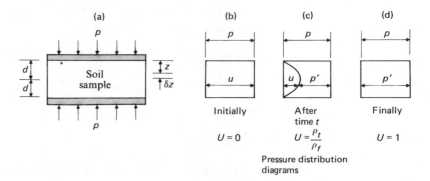

Figure 3.22

The time which this will take depends on:

(1) the distribution of effective pressure across the sample,
(2) the length of the drainage path d,
(3) whether the water drains from one or both surfaces,
(4) m_v for the soil,
(5) k for the soil.

The basic differential equation for one dimensional consolidation is:

$$\frac{\partial u}{\partial t} = \frac{k}{\gamma_w m_v} \times \frac{\partial^2 u}{\partial^2 z}$$

and the term $\dfrac{k}{\gamma_w m_v} = c_v$ is known as the *coefficient of consolidation*.

$$(3.7)$$

This equation has been solved for a number of common cases of effective pressure distribution in terms of two dimensionless numbers:

the *average degree of consolidation* $U = \dfrac{\text{consolidation after time } t}{\text{total consolidation}}$

$$= \frac{\rho_t}{\rho_f}$$

and a *time factor* $T_v = \dfrac{c_v \cdot t}{d^2}$ (3.8)

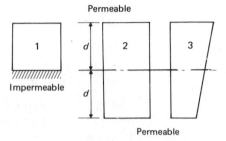

Permeable

Impermeable

Permeable

Figure 3.23

For the cases shown on Fig. 3.23:

$$U < 0.6: \quad T_v = \frac{\pi}{4} U^2 \tag{3.9}$$

$$U > 0.6: \quad T_v = [-0.933 \log_{10}(1 - U) - 0.0851] \tag{3.10}$$

Since $T_v \propto t$ any consolidation curves for a soil sample with the same pressure distribution and drainage conditions will be similar to the theoretical curves (Fig. 3.24).

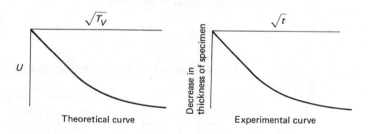

Figure 3.24

It may also be noted that the consolidation test provides an alternative method for determining the coefficient of permeability of a soil which would be too impermeable for normal permeameter tests. From eqn (3.7) $k = \gamma_w \cdot m_v \cdot c_v$ (care should be taken with units).

COMPRESSIBILITY OF SOIL AND SETTLEMENT OF FOUNDATIONS **77**

3.10 Coefficient of consolidation of a soil sample

The following time/settlement record was obtained from a consolidation test on a soft clay sample of average initial thickness 19 mm, for a given intensity of loading, drainage being through both top and bottom surfaces:

Time (min)	$\frac{1}{4}$	1	$2\frac{1}{4}$	4	9	16	25	36	49	24 h
Settlement $(10^{-3}$ mm)	140	282	424	564	839	1019	1148	1231	1257	1309

(a) Assuming that the equation of the initial portion of the degree of consolidation/$\sqrt{\text{time}}$ curve is

$$U = \frac{2}{\sqrt{\pi}} \sqrt{\frac{c_v \cdot t}{d^2}}$$

find the coefficient of consolidation c_v of the clay, neglecting the effect of secondary compression.
Estimate:

(b) the amount of 60% of the ultimate settlement of a bed of the tested clay, 1·5 m thick, when subjected to the same intensity of loading and the same conditions of drainage.

(c) the time required to reach 60% of the ultimate settlement of a bed of the same clay which lies on an impermeable stratum so that drainage is through the top surface only.

(d) the average value of the coefficient of volume decrease m_v in terms of the given intensity of loading p.   (UL)

Solution   The formula given is a rearrangement of the formulae for the time factor (3.8) and (3.9).

It is first necessary to calculate U. The final settlement ρ_{vf} after 24 h = 1 309 × 10^{-3} mm

$$\therefore U = \frac{\rho_{vt}}{1\,309 \times 10^{-3}}$$

The calculated values are tabulated below:

t (min)	$\frac{1}{4}$	1	$2\frac{1}{4}$	4	9	16	25	36	49	24 h
$\sqrt{t}$	$\frac{1}{2}$	1	$1\frac{1}{2}$	2	3	4	5	6	7	
U	0·107	0·215	0·324	0·431	0·640	0·780	0·876	0·940	0·960	1·00

From these figures Fig. 3.25 is plotted.
(a) From Fig. 3.25:

$$U = 0{\cdot}640 \quad \text{when} \quad t = 9 \text{ min} = 540 \text{ s}$$

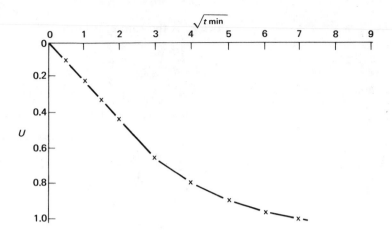

Figure 3.25

Substituting in the formula $U = \dfrac{2}{\sqrt{\pi}} \sqrt{\dfrac{c_v t}{d^2}}$

$$0\cdot 640 = \frac{2}{\sqrt{\pi}} \sqrt{\frac{c_v \cdot 540}{19^2}}$$

$$\therefore \ c_v = \mathbf{0\cdot 215 \ mm^2/s}$$

(*b*) When $U = 0\cdot 6$

Settlement $= 0\cdot 6 \times 1\,309 \times 10^{-3} = 785 \times 10^{-3}$ mm for sample 19 mm thick

$\therefore$ By simple proportion, for the same conditions of drainage and loading, a stratum $1\cdot 5$ m thick would be expected to settle an amount

$$= \frac{(1\cdot 5 \times 10^3) \times 785 \times 10^{-3}}{19} = \mathbf{62 \ mm.}$$

(*c*) When $U = 0\cdot 60$, $\sqrt{t} = 2\cdot 8$ $\therefore$ $t = 7\cdot 84$ min

For the soil sample, the length of the drainage path

$= \frac{1}{2}$ mean thickness of sample

$$= \frac{1}{2} \cdot \frac{19 + (19 - 785 \times 10^{-3})}{2} = 9\cdot 304 \text{ mm.}$$

For stratum of thickness z m, the length of the drainage path $= z$ m since the drainage can only occur through the top surface in the given circumstances.

Since c_v is the same for both:

$$\therefore \left(\sqrt{\frac{t}{d^2}}\right)_{\text{sample}} = \left(\sqrt{\frac{t}{d^2}}\right)_{\text{stratum}}$$

$$\therefore \frac{7\cdot84}{9\cdot304^2} = \frac{t}{(z \times 10^3)^2} \quad \therefore t = \frac{7\cdot84 \times 10^6}{9\cdot304^2} \cdot z^2 \min \left[\frac{\text{day}}{60 \times 24 \min}\right]$$

$$= 62\cdot9z^2 \text{ days}$$

(d) $$m_v = -\frac{1}{h} \cdot \frac{dh}{dp} \qquad\qquad (3.6)$$

At start $h = 19$ mm

At finish $dh = 1\,309 \times 10^{-3}$ mm

$$\therefore m_v = -\frac{1\,309 \times 10^{-3}}{19\,dp}$$

$$= -\frac{69 \times 10^{-3}}{dp}\frac{m^2}{kN}$$

3.11 Time for settlement of a foundation

A layer of compressible clay 6 m thick lies on an impervious bed of rock and carries an overburden of pervious sand. A large structure founded in the sand causes the pressure on every horizontal section of the clay to increase to the same value. In a standard laboratory oedometer test the void ratio of a sample of the clay (19 mm thick) decreased from 0·765 to 0·750 under a corresponding increase in pressure. Consolidation was 70% complete after 30 min.

Estimate the settlement of the structure and the time elapsing before one half of this amount has taken place. The appropriate time factor/degree of consolidation curve is defined by the following values:

U	0·40	0·60	0·80
T_v	0·13	0·28	0·57

(ICE)

Solution In solution 3.5 it was shown that in a consolidation test, since there is no change in the cross-section of the sample,

$$\frac{dh}{h} = \frac{de}{1+e_0}$$

$\therefore$ Change in thickness of sample during test

$$dh = 19 \times \frac{(0\cdot765 - 0\cdot750)}{1 + 0\cdot765}$$

$$= 0\cdot1615 \text{ mm}$$

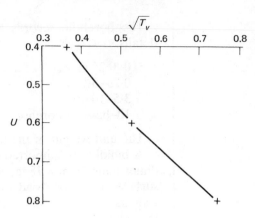

Figure 3.26

∴ By proportion for similar loading conditions, a stratum 6 m thick will settle

$$\rho_v = \frac{0 \cdot 1615 \times (6 \times 10^3)}{19}$$

$$= 51 \text{ mm}$$

Figure 3.26 shows a plot of the given values of $U/\sqrt{T_v}$. From the graph when $U = 0 \cdot 70$, $T_v = 0 \cdot 41$

$$T_v = \frac{c_v t}{d^2}$$

$$\therefore \ 0 \cdot 41 = \frac{c_v \times (30 \times 60)}{(19/2)^2}$$

$$\therefore \ c_v = \textbf{0·02 mm}^2\textbf{/s}$$

For 50% of total settlement of structure

$$U = 0 \cdot 50 \quad \therefore \ T_v = 0 \cdot 19$$

$$T_v = \frac{c_v \cdot t}{d^2}$$

$$0 \cdot 19 = \frac{0 \cdot 02 \times t}{(6 \times 10^3)^2}$$

$$\therefore \ t = \frac{0 \cdot 19 \times (6 \times 10^3)^2}{0 \cdot 02} \text{ s} \left[\frac{\text{year}}{60 \times 60 \times 24 \times 365 \text{ s}}\right]$$

$$= \textbf{10·8 years.}$$

3.12 Primary and secondary consolidation of a soil mass

Distinguish between primary and secondary consolidation, and describe a curve fitting method applicable to the results of a laboratory consolidation test.

The borehole records at a level site revealed the following details:

0–0·25 m	Topsoil.
0·25–3·5 m	Sand. Water table at 1·25 m.
3·5–10 m	Clay.
10–base of borehole	Impervious shale.

The unit weight of the topsoil and sand is $19·2 \text{ kN/m}^2$.

A building is supported on a partially compensated stiff raft of dimensions $18 \text{ m} \times 18 \text{ m}$, with its base at 2·25 m below ground surface. The gross load from the raft and building is $7 \times 10^4 \text{ kN}$. Calculate the settlement of the structure due to consolidation of the clay layer, whose coefficient of volume compressibility decreases linearly from $4·2 \times 10^{-4} \text{ m}^2/\text{kN}$ at its top to $2·7 \times 10^{-4} \text{ m}^2/\text{kN}$ at its base. Use the pressure influence chart Fig. 3.6.

Describe why the stress-history of the clay might lead you to amend the result of this calculation. How could the additional settlement due to compression of the sand be assessed?   (ICE)

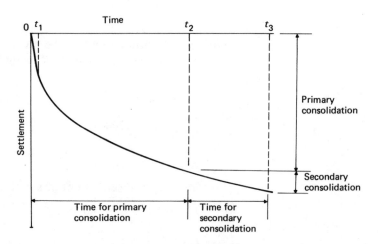

Figure 3.27

Solution   The elements of a time-settlement curve of a cohesive soil under a constant load are shown on Fig. 3.27. The time from 0 to t_1 when any air in the voids is expelled, and the time from t_1 to t_2 when water is expelled from the voids until the soil reaches a state of equilibrium, are the total time for the primary consolidation. This constitutes the greater part of the settlement under load.

Beyond time t_2 the process of consolidation continues slowly. The reasons for this are not well understood but they have to do with changes in the moisture films which surround the soil particles. The consolidation due to these effects is known as the secondary consolidation.

In order to separate the effects of secondary and primary consolidation in the results obtained from oedometer tests, Taylor devised a construction known as the 'square root of time' fitting method.

The theoretical relationship between U and T_v for the drainage conditions in an oedometer test are:

$$U < 0.6: \quad T_v = \frac{\pi}{4} U^2$$

$$U > 0.6: \quad T_v = [-0.933 \log_{10}(1 - U) - 0.0851]$$

If these are plotted on a graph of $U/\sqrt{T_v}$ the resulting curve is shown on Fig. 3.28.

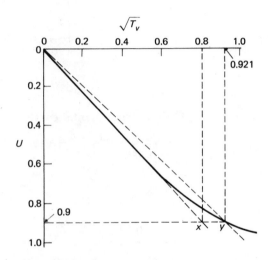

Figure 3.28

If the straight portion is produced to x where $U = 0.9$, $\sqrt{T_v} = 0.8$, the corresponding point y on the actual $\sqrt{T_v}$ curve is $\sqrt{T_v} = 0.921$. Thus a straight line drawn from 0 to y will have values of $\sqrt{T_v}: \dfrac{0.921}{0.8}$

$= 1.15$ times greater than corresponding values on line 0 to x.

Since in an oedometer test, the change in thickness of the sample is related to U and the time to T_v, a similar shaped curve is produced when the change in thickness of a sample is plotted against $\sqrt{t}$. Figure 3.29 shows a typical experimental curve.

To fit this to the theoretical curve, a line is drawn from 0 with $\sqrt{t}$ values 1.15 times those on the straight portion of the experimental curve. Where this straight line cuts the curve is taken to be the point corresponding with 90% primary settlement (i.e. $U = 0.9$). To find the change in thickness corresponding to $U = 1$, the change in thickness corresponding to 90% primary settlement is multiplied by $\frac{10}{9}$.

The change in thickness of the sample beyond this point is attributed to secondary consolidation.

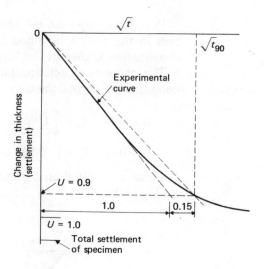

Figure 3.29

On the experimental curve

Where $U = 0.9$, $\sqrt{t} = \sqrt{t_{90}}$ and the corresponding theoretical value of $T_v = 0.848$

$$T_v = \frac{c_v t}{d^2}$$

$$\therefore \; 0.848 = \frac{c_v \cdot t_{90}}{d^2}$$

$$\therefore \; c_v = \frac{0.848 d^2}{t_{90}}. \tag{3.9}$$

For the given site:

Soil load removed in construction of raft

$$= 19.2 \times 18^2 \times 2.25 = 1.4 \times 10^4 \text{ kN}.$$

This represents the partial compensation of the raft load referred to in the question.

$\therefore$ Net load on soil due to structure $= (7 - 1.4) \times 10^4 = 5.6 \times 10^4 \text{ kN}$.

If this is assumed to be uniformly distributed under the raft, the contact pressure between the soil and the raft

$$q = \frac{5.6 \times 10^4}{18 \times 18} = 173 \text{ kN/m}^2.$$

(In solution 3.2 it was shown that the distribution of contact pressure under a stiff foundation is not uniform. For calculations of pressure distribution at some distance below the foundation, however, it is common to assume a uniform distribution.)

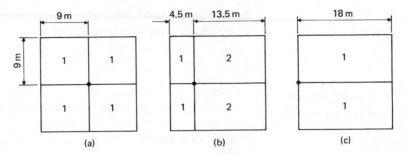

Figure 3.30

In order to estimate the settlement of the raft due to consolidation of the clay layer, the influence chart Fig. 3.6 may be used. If the raft is split up into areas shown on Fig. 3.30, the pressure at the centre of the clay layer may be found at the centre, quarter points and centre of the outer edge of the raft.

The depth from the base of the raft to the centre of the clay layer $z = 5 \cdot 50$ m. Then using the influence chart:

Fig. 3.30(a) Square 1 $B/z = L/z = 2$ $\therefore I = 0 \cdot 233$

and since all squares are the same

$$\therefore \sigma_z = 0 \cdot 233 \times 173 \times 4 = 161 \text{ kN/m}^2$$

Fig. 3.30(b) Square 1 $B/z = 1$, $L/z = 2$ $\therefore I = 0 \cdot 200$

Square 2 $B/z = 3$, $L/z = 2$ $\therefore I = 0 \cdot 238$

$$\therefore \sigma_z = 173(0 \cdot 200 \times 2 + 0 \cdot 238 \times 2) = 151 \text{ kN/m}^2$$

Fig. 3.30(c) Square 1 $B/z = 4$, $L/z = 2$ $\therefore I = 0 \cdot 240$

$$\therefore \sigma_z = 173 \times 0 \cdot 240 \times 2 = 83 \text{ kN/m}^2$$

$\therefore$ Average increase at centre of clay stratum due to construction of raft

$$= \frac{(161 + 151 + 83)}{3} = 132 \text{ kN/m}^2 \text{ approximately}$$

m_v at centre of clay layer $= \dfrac{(4 \cdot 2 + 2 \cdot 7)10^{-4}}{2} = 3 \cdot 45 \times 10^{-4} \text{ m}^2/\text{kN}$

$$\therefore \text{Settlement} = m_v \times p \times h$$
$$= 3 \cdot 45 \times 10^{-4} \times 132 \times 6 \cdot 5 \times 10^3 = \textbf{296 mm.}$$

If the clay were overconsolidated the settlement would be likely to be less than that calculated assuming it to be normally consolidated since the e/p curve tends to be a re-compression curve with a flatter slope rather than a compression curve. Thus the values of m_v tend to be smaller over corresponding pressure ranges.

The compression of the sand can be found from the results of standard penetration tests. These give an indication of the relative density of the sand from which an estimate of its likely settlement can be obtained using empirical charts.

3.13 Coefficient of consolidation of soil sample using $\sqrt{t}$ fitting method

The following results were obtained from a consolidation test on a sample of clay with drainage from both top and bottom surfaces when the load was increased from 100 kN/m² to 200 kN/m²:

Time t from application of pressure increment (min)	Decrease in thickness of sample (mm)
$\frac{1}{4}$	0·175
1	0·305
$2\frac{1}{4}$	0·432
4	0·558
$6\frac{1}{4}$	0·658
9	0·752
16	0·884
25	0·968
36	1·011
24 h	1·104

The initial thickness of the sample was 19·65 mm.

Plot the curve of decrease in thickness against $\sqrt{t}$.

Correct the curve using the 'square root of time' fitting method to allow for the effects of secondary consolidation and hence calculate the value of the coefficient of consolidation c_v assuming the formula for the straight portion of the graph

$$U = \frac{2}{d} \sqrt{\frac{c_v \cdot t}{\pi}}$$

where U = degree of consolidation and d = length of the drainage path.

(HNC)

Solution The curve of decrease in thickness $\sqrt{t}$ for the given figures is plotted on Fig. 3.31.

It can be seen that the first portion is slightly curved. This is probably due to the expulsion of air from the voids, and the true start of the curve is found by extending the straight portion backwards to x. From x a straight line is drawn xy with values of $\sqrt{t}$ 1·15 times the corresponding values of $\sqrt{t}$ on the straight portion of the experimental graph. Where the line xy cuts the original curve is taken as the amount of 90% of the primary settlement.

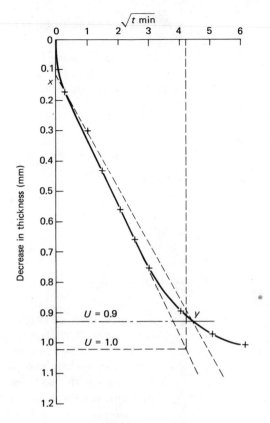

Figure 3.31

From the graph: $U = 0$ at decrease in thickness $= 0.12$ mm

$U = 0.9$ at decrease in thickness $= 0.93$ mm

$\therefore$ $U = 1.0$ at decrease in thickness

$$= (0.93 - 0.12) \times \tfrac{10}{9} + 0.12 = 1.02 \text{ mm.}$$

The corresponding value of $\sqrt{t}$ on the straight portion of the experimental curve (extended) $= 4.2$ and substituting in the given formula

$$U = \frac{2}{d} \sqrt{\frac{c_v t}{\pi}}$$

$U = 1.0$ $\sqrt{t} = 4.2$

$d = \tfrac{1}{2}$ average thickness of sample from $U = 0$ to $U = 1$

$$= \frac{1}{2}\left(19.65 - \frac{1.02}{2}\right) = 9.57 \text{ mm}$$

$$\therefore \ 1 = \frac{2}{9.57} \times 4.2 \sqrt{\frac{c_v}{\pi}}$$

$\therefore$ $c_v = 4.08 \text{ mm}^2/\text{min} = \mathbf{0.068 \ mm^2/s.}$

An alternative solution is to use the value $U = 0.9$. Then

$$\sqrt{t} = 4.45 \qquad d = \frac{1}{2}\left(19.65 - \frac{0.93}{2}\right) = 9.59$$

and from eqn (3.9)

$$c_v = \frac{0.848d^2}{t_{90}}$$

$$= \frac{0.848 \times 9.59^2}{4.45^2} = 3.94 \text{ mm}^2/\text{min} = \mathbf{0.066 \text{ mm}^2/\text{s}.}$$

In general at any particular stress an overconsolidated clay is less compressible and has a higher shear resistance than a normally consolidated clay.

Problems

1. The centres of two columns A and B are 3 m apart. Column A is supported by a square footing $1.2 \text{ m} \times 1.2 \text{ m}$, the base of which is 2.5 m below ground level. The footing of column B is $1 \text{ m} \times 1 \text{ m}$ and its base is 1.5 m below ground level. The contact pressure on the soil under each footing is 425 kN/m^2. Find the pressure on the soil per horizontal square metre (additional to the original overburden pressure) at a depth of 9 m below ground level (*a*) vertically below the centres of A and B; (*b*) midway between them.

The pressure at depth z and horizontal distance r from the line of action of a concentrated load P is $\dfrac{KP}{z^2}$, where K is an influence factor, values of which are as follows:

r/z	K
0	0.478
0.1	0.466
0.2	0.433
0.3	0.385
0.4	0.329
0.5	0.273

(ICE)

$[(a)\ 9.41 \text{ kN/m}^2,\ 8.04 \text{ kN/m}^2;\ (b)\ 9.34 \text{ kN/m}^2]$

2. Two column loads of P and $2P$ respectively are to be applied to the surface of a 6 m thick layer of dense sand which overlies a layer of clay. Determine the maximum spacing of the columns if the settlement of the heavier column is not to be greater than 1.5 times that of the other column. Assume the settlement is due to the clay alone and that its compressibility characteristics are the same for each load.

(5·16 m)

3. During a consolidation test an increment of pressure of 215 kN/m² was applied to a sample of saturated clay initially 23·6 mm · thick. When equilibrium was reached the thickness of the sample was reduced to 21·9 mm. The pressure was then removed and the sample allowed to expand and take up water. The final thickness was 22·9 mm and the final water content 26·8%. If the specific gravity of the particles was 2·69 and the sample was saturated throughout the test, find the void ratio before and after the consolidation.

Find also the average coefficient of volume decrease (decrease in unit volume/unit increase in pressure) over the range of pressure applied. (ICE)

(0·77; 0·64; 0·335 × 10⁻³ m²/kN)

4. The following information was obtained from a borehole (levels in m below ground surface):

> 0 to −4   fine, sand, dry density 1·52 Mg/m³,
> specific gravity of particles 2·65
> −4 to −7·5   stiff clay, wet density 1·98 Mg/m³
> below −7·5   medium sand.

Calculate the effective pressure at the top and bottom surface of the clay stratum:
   (*a*) when the water table is at the surface,
   (*b*) when the water table has been lowered 2·5 m by pumping. (It may be assumed that the sand above the water table remains saturated with held water.)

If the average value of the coefficient of compressibility of the clay is 0·56 × 10⁻³ m²/kN, find the approximate reduction in the thickness of the clay due to consolidation as a result of lowering the water table. (ICE)

[(*a*) 37·84 kN/m², 72·14 kN/m²; (*b*) 25 kN/m² increase; 0·049 m]

5. The foundation of a large building is at 3 m below ground level. The soil profile is as follows:

Strata	Depths from ground surface (m)	Average density (Mg/m³)
Topsoil and sand	0 to −7·5 (water table at −6·75)	2·00
Clay	−7·5 to −10·5	1·92
Impervious rock	below −10·5	

The average coefficient of compressibility of the clay is $177 \times 10^{-6} \, \text{m}^2/\text{kN}$ and the coefficient of consolidation is $0 \cdot 025 \, \text{mm}^2/\text{s}$. The additional pressure on the clay due to the weight of the building is $200 \, \text{kN/m}^2$ at the top of the stratum, decreasing uniformly with depth to $150 \, \text{kN/m}^2$ at the bottom. Find:

(a) the original effective pressures at top and bottom of the clay stratum before excavation commences;

(b) the effective pressures at the top and bottom of the clay stratum after completion of the building;

(c) the magnitude of settlement of the building expected due to consolidation of the clay;

(d) the probable time in which 90% of this settlement will occur, taking the time factor as $0 \cdot 80$. (UL)

[(a) 140, 167·6 kN/m²; (b) 280, 257·6 kN/m²; (c) 0·061 m; (d) 9·1 years]

6. Below the foundation of a structure there is a stratum of compressible clay 6 m thick with incompressible porous strata above and below. The average overburden pressure on the stratum before construction was $115 \, \text{kN/m}^2$ and after completion of the structure the pressure increased to $210 \, \text{kN/m}^2$. Oedometer tests were carried out on a sample of clay initially 20 mm thick. Each pressure was allowed to act for 24 h and the decrease in thickness measured, the results being as follows:

Pressure (kN/m²)	Thickness (mm)
0·0	20·0
50	19·8
100	19·4
200	19·0
400	18·6

Under a pressure of $100 \, \text{kN/m}^2$, 90% of the total consolidation took place in 21 min.

Find (a) the probable settlement of the structure, and (b) the time in which 90% of this settlement may be expected to occur. (ICE)

[(a) 0·112 m; (b) 3·8 years]

7. Under a large building there is compact sand of density $2 \cdot 08 \, \text{Mg/m}^3$ extending for a depth of 3 m below the foundation. Under this there is a stratum of clay of density $1 \cdot 65 \, \text{Mg/m}^3$, 1 m thick and beneath the clay there is more sand. The water table is at foundation level. The following results were obtained from a

consolidation test on a sample of the clay:

Pressure (kN/m²)	Thickness of sample (mm)
0·0	19·0
25·0	18·8
50·0	18·3
100·0	17·6
200·0	16·1
400·0	14·5

If the mean value of the additional pressure on the clay from the weight of the building is $150 \, kN/m^2$ estimate the probable settlement due to the consolidation of the clay.   (ICE)

(0·126 m approx.)

8. A large foundation slab is supported on a bed of compact sand which extends to a depth of 6 m below the base. Under the sand there is a stratum of clay of thickness 4·8 m which in turn rests on impermeable shale. The initial overburden pressure at the top of the clay is $85 \, kN/m^2$. The additional pressures applied by the foundation load are as follows:

	Pressure (kN/m²)	
	under centre of slab	under corner of slab
Top of clay stratum	85	45
Bottom of clay stratum	25	105

The clay has an average density of $1·92 \, Mg/m^3$ and oedometer tests gave the following data:

Effective pressure (kN/m²)	Void ratio
50	0·93
100	0·91
200	0·88
400	0·85

Estimate the final settlement under the centre and under a corner of the foundation slab due to consolidation of the clay stratum. Ground water level is situated within the sand.   (ICE)

(0·037 m, 0·051 m)

9. The following readings were obtained during one stage of a consolidation test on a sample of saturated clay, initially 190 mm thick, with drainage from both top and bottom faces of the sample:

Time (min)	Reduction in thickness (mm × 10⁻³)	Time (min)	Reduction in thickness (mm × 10⁻³)
0	0	9	432
0·25	109	16	526
1	173	25	610
2·25	236	36	665
4	302	49	690
6·25	355	24 h	770

Plot the curve of reduction in thickness against square root of time and from it determine the coefficient of consolidation of the clay in m^2/day.

Explain briefly what is meant by *secondary consolidation*.

(ICE)

$(c_v = 6·46 \times 10^{-3}\, m^2/\text{day})$

10. On a flat site boreholes reveal that a 4 m layer of dense sand overlies a layer of compressible clay 8 m thick, below which stiffer clay extends to a considerable depth.

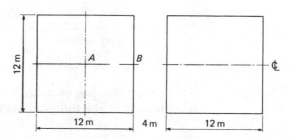

Figure 3.32

It is proposed to store material on two adjacent paved areas as indicated in Fig. 3.32. The pavement and the stored material together can be assumed to provide a uniform contact pressure q on the surface of the ground. Estimate the value of q in kN/m^2 if the differential settlement between points A and B is to be limited to 0·05 m. The appropriate values of the coefficient of compressibility for the 8 m clay layer are:

upper 4 m, $m_v = 0·746 \times 10^{-6}\, m^2/N$
lower 4 m, $m_v = 0·559 \times 10^{-6}\, m^2/N$

Use pressure influence chart Fig. 3.6.

(ICE)

$(10\, kN/m^2)$

4

The shearing resistance of soil

The strength of a soil depends on its resistance to shearing stresses. It is made up basically of two components:

(1) frictional—due to interlocking and friction between individual particles;
(2) cohesive—due to adhesion between the soil particles.

The two components are combined in Coulomb's shear strength equation:

$$\tau_f = c + \sigma \tan \phi \qquad (4.1)$$

where τ_f = shearing resistance of soil at failure = shear strength

c = apparent cohesion of soil

σ = total normal stress on shear plane

ϕ = angle of shearing resistance of soil

The cohesive resistance c is assumed to be constant but the frictional resistance increases with an increase in the normal stress σ. Thus eqn (4.1) represents a straight line and if σ is plotted against τ_f the result shown in Fig. 4.1 is obtained.

Worked examples

4.1 Shearing resistance of a soil sample using a shear box test

The following readings were taken in a shear box test on compacted sand:

Normal load (N):	90	225	350
Shear load (N) (peak):	85	215	343
Shear load (N) (ultimate):	55	143	230

The shear box measured 60 mm square.

Find the angle of shearing resistance (*a*) in the compacted state; (*b*) in the loose state, and estimate the probable angle of repose of this material. (ICE)

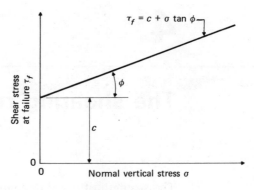

Figure 4.1

Solution The shear box test is normally used to determine the shearing resistance of a cohesionless soil. The apparatus is shown diagrammatically in Fig. 4.2.

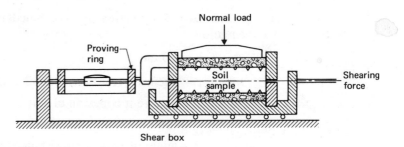

Figure 4.2

The box is 60 mm square, open at the top and bottom, and divided horizontally into two halves. It is placed in a rectangular container designed to hold the bottom half of the box rigidly in position. A porous stone is placed in the bottom of the box, followed by a perforated, toothed metal plate. The soil sample is then placed in the box with another plate and porous stone on top. Above the top porous stone a pressure pad is placed.

The test is performed by applying a vertical load to the sample and then applying a horizontal force at a constant rate of strain until the sample fails in shear, i.e. the particles begin to move relative to one another. The horizontal shearing force is measured by means of a proving ring.

This procedure is repeated on similar samples of soil with different vertical loads.

From the results obtained a graph is plotted of shearing force/normal vertical load or shearing stress (at failure)/vertical stress. A straight line drawn through the points obtained is taken to represent the Coulomb equation and the values of c and ϕ found from the graph (Fig. 4.1). (It should be noted that if measurements are taken off the graph the same scale must be used on both axes.)

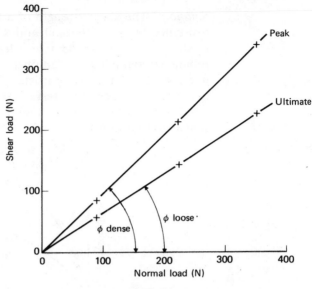

Figure 4.3

Figure 4.3 has been plotted from the figures given. From the graph it is found that:

(*a*) in the compacted state (peak value) $\phi = $ **44°**;
(*b*) in the loose state $\phi = $ **33°**.

The value of the apparent cohesion $c = 0$ in both cases as would be expected for a cohesionless soil.

The angle of repose is the angle to the horizontal at which a heap of dry material will stand without support. It will be about the same in this case as the angle of shearing resistance in the loose state, i.e. **33°**.

4.2 Shearing resistance of a compacted sand

Make a sketch of the shearing force/deformation curves you would expect from shear box tests on (*a*) loose sand; (*b*) compacted sand.

Shear box tests on compacted sand gave the following results:

Normal pressure (kN/m²)	Shearing stress (kN/m²)	
	Peak value	Ultimate value
35	29	23
70	58	45
105	87	67

Find the angle of shearing resistance of the sand (*a*) in the compacted state; (*b*) when loosened by the shearing action.
(ICE)

Solution The shear strength of a loose sand consists of the frictional resistance between the soil grains. In a dense sand or a compacted sand (i.e. a sand made more dense by mechanical means such as rolling or ramming), additional shear strength is provided by the interlocking of the soil particles. This accounts for the peak values obtained in a shear box test. If the strain continues to increase, the interlocking is broken down and the shearing resistance falls to a value similar to that of a loose sand.

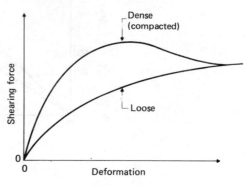

Figure 4.4

Typical shearing force/deformation curves for a loose and dense sand are shown on Fig. 4.4.

The results of the shear box test are plotted on Fig. 4.5.

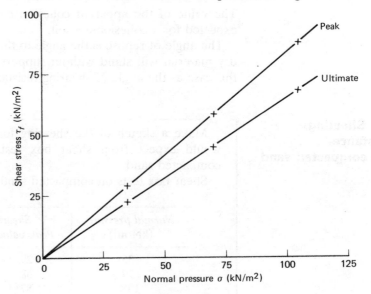

Figure 4.5

It can be seen from this that:

(*a*) in the compacted state $\phi = 40°$;
(*b*) in the loose state $\phi = 33°$.

Why does compaction alter the angle of shearing resistance of sand?

The following readings were taken using a standard shear box 60 mm × 60 mm in area containing a specimen of dry sand:

Time (min)	Proving ring dial* (mm × 10^{-3})	Vertical displacement dial (mm × 10^{-3})
0	0	0
2	323	−30
4	516	−20
6	582	+140
8	552	+348
10	478	+437
12	465	+462

* 1 div. = 0·5 N.

The total load from the hanger was 355 N and the screw jack moved the lower box at 0·5 mm/min. Plot the shearing resistance and vertical movement against displacement and calculate the angle of shearing resistance ϕ. Estimate the probable porosity of the sand (i.e. loose or dense) and the value of ϕ for such a porosity that no change of volume occurs during shear.

(ULKC)

Solution Compaction is the packing together of the soil particles by mechanical means with a resulting expulsion of air from the voids. The closer packing, thus brought about, tends to increase the interlocking resistance of the sand particles and hence increase the value of ϕ.

The horizontal displacement of the soil will be the movement of the lower half of the shear box (0·5 mm/min) minus the movement of the proving ring. The shearing resistance is the proving ring dual gauge reading × 0·5 N. Thus from the information given:

Time (min)	Movement of box (mm)	Movement of dial gauge (mm)	Displacement of soil (mm)	Shearing resistance (N)	Vertical movement (mm)
0	0	0	0	0	0
2	1	0·323	0·677	161	−0·030
4	2	0·516	1·484	258	−0·020
6	3	0·582	2·418	291	0·140
8	4	0·552	3·448	276	0·348
10	5	0·478	4·522	239	0·437
12	6	0·465	5·535	232	0·462

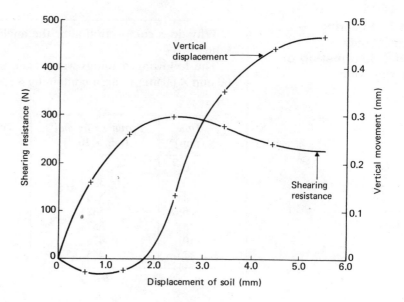

Figure 4.6

The values of vertical displacement and shearing resistance are plotted against deformation in Fig. 4.6.

The value of ϕ for no change in volume during shear may be estimated using the ultimate shearing resistance = 232 N.

Then using the equation $\tau_f = c + \sigma \tan \phi$

In the case of dry sand $c = 0$

$$\sigma = \frac{\text{vertical load}}{\text{area of sample}} = \frac{355}{A}$$

$$\tau_f = \frac{\text{ultimate shearing resistance}}{\text{area of sample}} = \frac{232}{A}$$

$$\therefore \tan \phi = \frac{232}{355} = 0 \cdot 653$$

$$\therefore \phi = \mathbf{33°}$$

Initially a dense sand tends to increase in volume during shear and a loose sand to decrease in volume. From the shape of the two curves therefore the soil behaves in the way a dense sand would be expected to react (Fig. 4.4).

For a dense sand, a representative value for the porosity $n = 0 \cdot 34$. (For a loose sand $n = 0 \cdot 45$.)

4.4 Comparison of shear resistance tests on a sand sample

These readings were taken during two drained direct shear box tests on the same sand:

Test 1

Relative horizontal displacement (mm $\times 10^{-3}$)	0	625	1 875	3 125	4 375	5 625
Vertical displacement of piston (+ve upwards) (mm $\times 10^{-3}$)	0	−89	−227	−275	−292	−295
Shear stress (kN/m^2)	0	58	75	84	87	87

Test 2

Relative horizontal displacement (mm $\times 10^{-3}$)	0	625	1 250	1 875	2 500	3 125	3 750	5 500	6 250
Vertical displacement of piston (+ve upwards) (mm $\times 10^{-3}$)	0	−8	43	152	251	373	480	559	569
Shear stress (kN/m^2)	0	91	108	113	116	116	112	100	89

The constant normal stress applied by the piston was 150 kN/m^2 in both tests.

Present the readings graphically and calculate the angle of shearing resistance for each test. Explain the differences you see and briefly comment on their significance in relation to the loads imposed during an earthquake. (CEI)

Solution The given values are plotted on Fig. 4.7. The density at which the sand neither expands nor contracts during shearing is known as the *critical density*. Sands with a density above this critical density may be regarded as dense whilst sands with a density below the critical are loose.

The results of test 1 indicate a loose sand since the sample compacts continuously under the shearing load. The results of test 2 with a peak shear stress which then drops off together with increasing vertical displacement indicate a dense sand.

Earthquakes set up vibrations which could cause a loose sand in the condition of test 1 to settle noticeably. The sand in the condition of test 2 is less likely to be seriously affected.

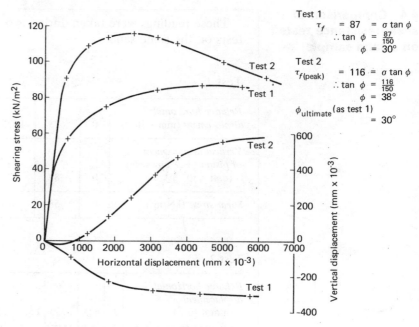

Test 1
$$\tau_f = 87 = \sigma \tan \phi$$
$$\therefore \tan \phi = \frac{87}{150}$$
$$\phi = 30°$$

Test 2
$$\tau_{f(peak)} = 116 = \sigma \tan \phi$$
$$\therefore \tan \phi = \frac{116}{150}$$
$$\phi = 38°$$

$\phi_{ultimate}$ (as test 1)
$$= 30°$$

Figure 4.7

4.5 Shearing resistance of a soil sample using a triaxial compression test

(a) Describe the triaxial compression test for the determination of the shearing resistance of soils, explaining the principles of three types of test which are commonly conducted.

(b) From a study of the Mohr circle diagram for a specimen tested in drained triaxial compression, show that at failure:

$$\sigma_1 = \sigma_3 \frac{1 + \sin \phi_d}{1 - \sin \phi_d} + \frac{2c_d \cdot \cos \phi_d}{1 - \sin \phi_d}$$

where σ_1 and σ_3 are the major and minor principal stresses.
(ICE)

Solution The triaxial compression apparatus is shown diagrammatically in Fig. 4.8.

The soil sample is cylindrical with a height to diameter ratio of 2:1. The sample is enclosed in a rubber membrane with porous stones at each end and placed in the perspex cell. The cell is sealed up and water pumped in to fill the cell at any required pressure. Thus initially the sample is subjected to a principal stress* σ_3 in all directions.

---

* A plane on which there is no shearing stress is known as a principal plane and the direct stress acting on the plane is known as a principal stress.

From the results obtained a series of Mohr's circles of stress are drawn. A line tangential to all the circles is drawn to represent the Coulomb shear strength equation and the appropriate values of c and ϕ are read off the resulting diagram.

When a stress is applied to a point P in the soil mass, the stress conditions on any

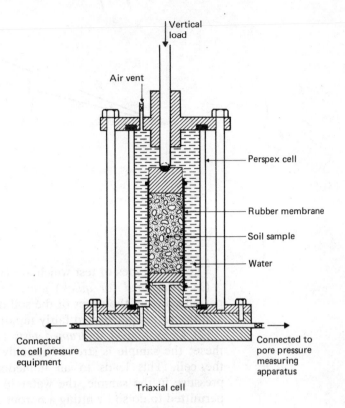

Vertical
load

Air vent

Perspex cell

Rubber membrane

Soil sample

Water

Connected
to cell pressure
equipment

Connected to
pore pressure
measuring
apparatus

Triaxial cell

Figure 4.8

A vertical load is applied through a proving ring at a constant rate of strain until the soil sample fails in shear. The total vertical stress on the sample is then σ_1 (the major principal stress) and since the sample was initially subjected to a stress σ_3 from the water in the cell, the additional vertical stress applied via the proving ring $= \sigma_1 - \sigma_3$ and this is referred to as the *deviator stress*.

The test is performed a number of times on similar samples using different initial cell pressures.

plane passing through the point will be made up of a direct stress acting at right angles to the plane and a shear stress tangential to the plane. There will be two planes on which no shear stresses occur and the direct stresses on these planes are the *major* and *minor* principal stresses.

These stress conditions can be represented graphically by a circle known as a *Mohr* circle (Fig. 4.9). It is drawn by plotting the major and minor principal stresses on the Ox axis and constructing a circle between them. The Oy axis represents the shear stress.

For any plane passing through P at an angle θ to the major principal plane, OX represents the normal stress on the plane and XY the shear stress. XOY represents the angle at which the resultant of these two stresses acts on the plane.

In the limiting case, where the soil will fail in shear, OY' is tangential to the circle and $X'OY' = \phi$.

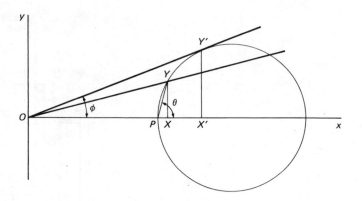

Figure 4.9

The three types of test which are commonly conducted are:

(1) *Undrained* (*or quick*) *tests*. In these, no drainage of water is permitted from the pores of the soil during the stressing of the sample and the test is performed fairly rapidly.

(2) *Consolidated undrained tests* (or consolidated quick tests). In these, the sample is stressed initially by the pressure of the water in the cell. This leads to an immediate increase in the pore water pressure in the sample, the water in the pores tries to escape and is permitted to do so by fitting a porous disc at the base thus allowing the sample to consolidate. When the pore water pressure returns to zero, the sample is sheared by increasing the vertical load, no further drainage being permitted.

(3) *Drained* (*or slow tests*). The sample is loaded and allowed to drain as in the previous test. The vertical load is then applied at such a slow rate that no pore water pressure develops in the sample.

(b) The Mohr circle diagram is shown on Fig. 4.10 together with the Coulomb line. This line is extended backwards to meet the horizontal axis at x.

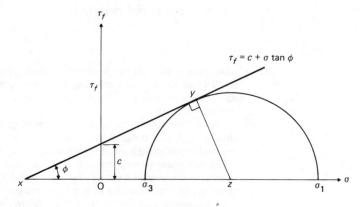

Figure 4.10

From the geometry of the figure:

$$\frac{xO + \sigma_1}{xO + \sigma_3} = \frac{xz + zy}{xz - zy} = \frac{1 + \dfrac{zy}{xz}}{1 - \dfrac{zy}{xz}} = \frac{1 + \sin\phi}{1 - \sin\phi}$$

$$\therefore \ xO + \sigma_1 = (xO + \sigma_3)\frac{1 + \sin\phi}{1 - \sin\phi} \quad \text{but} \quad xO = \frac{c}{\tan\phi}$$

$$\therefore \ \frac{c}{\tan\phi} + \sigma_1 = \frac{c}{\tan\phi}\left(\frac{1 + \sin\phi}{1 - \sin\phi}\right) + \sigma_3\left(\frac{1 + \sin\phi}{1 - \sin\phi}\right)$$

$$\therefore \ \sigma_1 = \sigma_3\left(\frac{1 + \sin\phi}{1 - \sin\phi}\right) + \frac{c}{\tan\phi}\left(\frac{1 + \sin\phi}{1 - \sin\phi} - 1\right)$$

$$\therefore \ \sigma_1 = \sigma_3\left(\frac{1 + \sin\phi}{1 - \sin\phi}\right) + \frac{2c\sin\phi}{\tan\phi(1 - \sin\phi)}$$

$$\therefore \ \boldsymbol{\sigma_1 = \sigma_3\left(\frac{1 + \sin\phi}{1 - \sin\phi}\right) + \frac{2c\cos\phi}{1 - \sin\phi}} \qquad (4.2)$$

4.6 Shearing resistance of a soil sample using undrained triaxial test

Explain the term *effective stress*.

Pore-pressure measurements were made during undrained triaxial tests on samples of compacted fill material from an earth dam. The results were as follows:

Lateral pressure	σ_3 kN/m^2	150	450
Total vertical pressure	σ_1 kN/m^2	400	1 000
Pore water pressure	u kN/m^2	+30	+125

Determine the apparent cohesion and the angle of shearing resistance (*a*) referred to total stress; and (*b*) referred to effective stress. (UL)

Solution The effective stress is the total stress minus the pore water pressure, i.e. $\sigma' = \sigma - u$. This concept was explained in Chapter 2.

The results given are used to plot two Mohr's circle diagrams and a line tangential to them is drawn to represent the Coulomb shear strength equation.

(*a*) Total stress circle 1: $\sigma_1 = \ 400$ kN/m^2 $\qquad \sigma_3 = 150$ kN/m^2

$\qquad\qquad\qquad$ circle 2: $\sigma_1 = 1\,000$ kN/m^2 $\qquad \sigma_3 = 450$ kN/m^2

These are plotted on Fig. 4.11, from which $c_u = \textbf{36 kN/m}^2$ $\phi_u = \textbf{20°}$

(*b*) To obtain the effective stress values, the pore water pressures are subtracted from the total pressures and a new Mohr's circle diagram plotted.

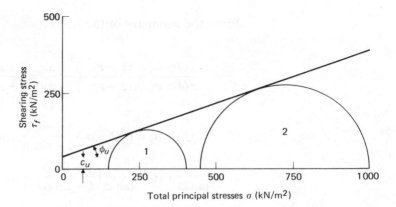

Figure 4.11

Effective stress circle 1: $\sigma_1' = 400 - 30 = 370 \text{ kN/m}^2$
$$\sigma_3' = 150 - 30 = 120 \text{ kN/m}^2$$
circle 2: $\sigma_1' = 1\,000 - 125 = 875 \text{ kN/m}^2$
$$\sigma_3 = 450 - 125 = 325 \text{ kN/m}^2$$

These are plotted on Fig. 4.12, from which

$$c' = \mathbf{24 \, kN/m^2} \qquad \phi' = \mathbf{25°}$$

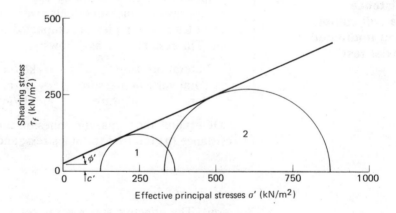

Figure 4.12

4.7 Comparison of shearing resistance tests to simulate field conditions

Enumerate the types of laboratory triaxial test you would specify to be carried out in connection with the following field problems:

(*a*) the stability of a clay foundation of an embankment; the rate of construction being such that some consolidation of the clay occurs;

(*b*) the initial stability of a footing on saturated clay;

(*c*) the long term stability of a slope in stiff fissured clay.

Give your reasons for your choice of test.

The table contains data obtained from *consolidated-undrained* tests on a soft clay. Determine the shear strength parameters in terms of effective stress. A different specimen of the same soil is tested in *undrained* triaxial compression at a cell pressure of $150 \, kN/m^2$, and fails when the deviator stress is $75 \, kN/m^2$. Calculate the pore pressure in the specimen at failure.

Cell pressure during consolidation and shear (kN/m^2)	At failure	
	Deviator stress (kN/m^2)	Pore water pressure (kN/m^2)
200	117	110
400	242	227
800	468	455

(ICE)

Solution The conditions reproduced during a laboratory triaxial test should be the same as those anticipated in the field for the particular investigation under consideration.

(*a*) Since there is some consolidation during construction of the embankment, a consolidated-undrained triaxial test with pore water pressure measurements would be appropriate in this case.

(*b*) A footing on saturated clay will initially increase the pore water pressure of the clay and only gradually, as consolidation occurs, will the effective stresses increase. The appropriate test in this case, therefore, would be an undrained triaxial test.

(*c*) The long term stability of a slope in a stiff fissured clay would depend on the effects of consolidation and water seepage. A drained test would give the necessary information about the long term shearing resistance of the clay. A consolidated-undrained test with pore water pressure measurements would also give the required information.

The shear strength parameters required are c' and ϕ'
The cell pressure $= \sigma_3$
The deviator stress $= \sigma_1 - \sigma_3$
$\therefore \sigma_1 =$ deviator stress $+ \sigma_3$.

The effective stresses are found by subtracting the corresponding pore water pressures.

Thus:

Test 1: $\sigma_1' = 200 + 117 - 110 = 207 \, kN/m^2$

$\sigma_3' = 200 - 110 \quad\quad = 90 \, kN/m^2$

Test 2: $\sigma_1' = 400 + 242 - 227 = 415 \, kN/m^2$

$\sigma_3' = 400 - 227 \quad\quad = 173 \, kN/m^2$

Test 3: $\sigma_1' = 800 + 468 - 455 = 813 \, kN/m^2$

$\sigma_3' = 800 - 455 \quad\quad = 345 \, kN/m^2$

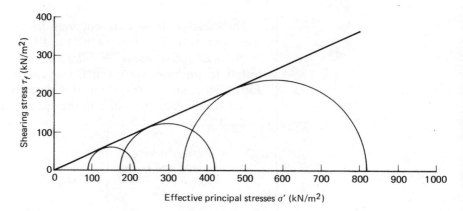

Figure 4.13

These values are used to plot Fig. 4.13 from which

$$c' = 0 \text{ kN/m}^2 \qquad \phi' = 25°$$

Using these values, Fig. 4.14 shows diagrammatically the Mohr's circle for the undrained test in terms of effective stresses where $\sigma_1 = 150 + 75 = 225 \text{ kN/m}^2$ and $\sigma_3 = 150 \text{ kN/m}^2$.

Then $x = 37.5 \text{ cosec } 25 = 37.5 \times 2.37 = 88.7$

$$\sigma_3 - u = 88.7 - 37.5 = 51.2 \text{ kN/m}^2$$

and $\qquad \sigma_3 = 150 \text{ kN/m}^2$

$$\therefore \ u = 150 - 51.2 = \textbf{98.8 kN/m}^2$$

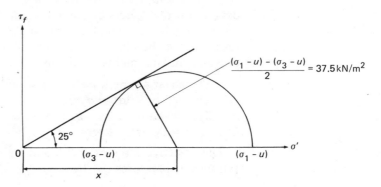

Figure 4.14

4.8 Pore-water pressure measurement in triaxial test

Describe briefly the method of measuring the pore water pressure in an undrained triaxial test.

A specimen of saturated sand 75 mm long by 37.5 mm diameter was fully consolidated in a triaxial cell at a cell pressure of 70 kN/m². The cell pressure was then raised to 550 kN/m² and an undrained triaxial test on the same sample produced the following results:

Decrease in length (mm)	Proving ring reading*	Cell pressure (kN/m²)	Pore water pressure (kN/m²)
0	0	550	480
6·25	435	550	348
12·5	813	550	202
18·75	1 020	550	85
25·0	1 140	550	51

* 1 div. = 1·5 N

Plot $(\sigma_1' - \sigma_3')$ and (σ_1'/σ_3') against axial strain and the Mohr circle at failure and comment on the result. (HNC)

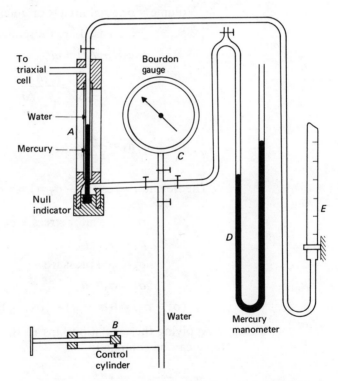

Pore water pressure measuring appartus (after Bishop)

Figure 4.15

Solution A sketch of the pore water pressure measuring apparatus devised by Bishop is shown in Fig. 4.15.

The pore-pressure measuring connection on the triaxial cell is connected to a null indicator A by a pipe filled with water. The null indicator consists essentially of a glass capillary tube inserted into an enclosed trough containing mercury. An increase in the pore water pressure in the soil in the cell will tend to lower the mercury in the capillary tube. In an undrained test no water can be allowed to escape from the soil, so the mercury is brought back to its original level by screwing in the piston in the control cylinder B. The pressure required to do this will be the pore water pressure and can be measured on the Bourdon pressure gauge C or for negative pressures on the mercury manometer D. The burette E is used to determine the pressure gauge and manometer readings at zero pore pressure.

Since the sample tested is saturated and the test is undrained, there will be no change in the volume of the sample. Thus if the length of the sample is reduced, the cross-sectional area must increase.

$$\text{The axial strain } \varepsilon = \frac{\text{change in length}}{\text{original length}} = \frac{\Delta l}{l}$$

Volume = original area $\times$ original length = $a \times l$

$\quad$ = increased area $\times$ reduced length = $a_1 \times (l - \Delta l)$

$\therefore \; a_1(l - \Delta l) = al$

$\therefore \; a_1 = \dfrac{a \cdot l}{l - \Delta l} = \dfrac{a}{1 - \dfrac{\Delta l}{l}}$

$\therefore \; a_1 = \dfrac{a}{1 - \varepsilon}$

$\therefore$ The deviator stress $\sigma_1 - \sigma_3 = $ load on proving ring $\times \dfrac{(1 - \varepsilon)}{a}$

$\quad \sigma_1 = $ deviator stress $+ \sigma_3$

$\quad \sigma_1' = \sigma_1 - u$

$\quad \sigma_3 = $ cell pressure

$\quad \sigma_3' = \sigma_3 - u$

$\therefore (\sigma_1' - \sigma_3') = [(\sigma_1 - u) - (\sigma_3 - u)] = \sigma_1 - \sigma_3.$

Applying these relationships to the given information the following table can be compiled:

Δl (mm)	ε	$\sigma_1' - \sigma_3'$ (kN/m^2)	σ' (kN/m^2)	σ_3' (kN/m^2)	σ_1'/σ_3'
0	0	0	70	70	1·0
6·25	0·0833	542	744	202	3·66
12·5	0·1667	920	1 268	348	3·62
18·75	0·2500	1 039	1 504	465	3·24
25·00	0·3333	1 031	1 530	449	3·07

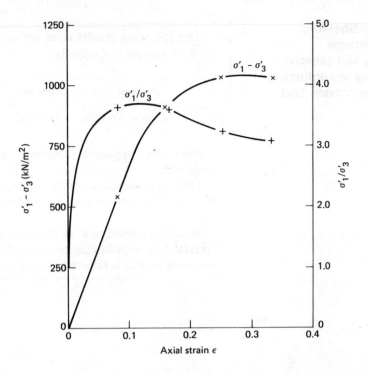

Figure 4.16

The required graphs are plotted on Fig. 4.16. The shapes of the curves indicate that the soil was overconsolidated.

At failure $\sigma_1' = 1\,530 \text{ kN/m}^2$ and $\sigma_3' = 449 \text{ kN/m}^2$.

These values are used to plot the Mohr circle (Fig. 4.17).

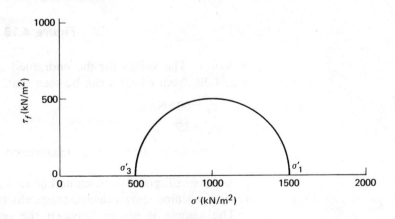

Figure 4.17

4.9 Shearing resistance of a soil sample using unconfined compression test

The following results were obtained from undrained shear-box tests on samples of silty clay:

Normal pressure (kN/m²)	200	300	400
Shear strength (kN/m²)	113	141	167

Find the apparent cohesion and the angle of shearing resistance.

Find also the value of the apparent cohesion which would be expected from an unconfined compression test on a sample of the same soil.

If another specimen of this soil is subjected to an undrained triaxial test with lateral pressure 275 kN/m², find the total axial pressure at which failure would be expected.　　　　(ICE)

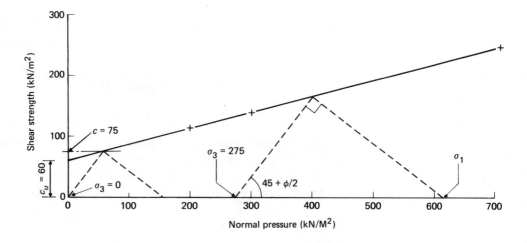

Figure 4.18

Solution　The values for the undrained shear-box test are plotted on Fig. 4.18, from which it can be seen that:

$$c_u = 60 \text{ kN/m}^2$$

$$\phi_u = 15°.$$

It is possible to perform an unconfined compression test in a triaxial testing machine by keeping the cell pressure at zero. It is more usual to use a small, portable machine known as the unconfined compression testing machine shown diagrammatically on Fig. 4.19.

The sample is placed between the central fixed plate and lower moving plate. By turning the handle, the upper moving plate is raised and this extends the spring. The force in the spring is proportional to

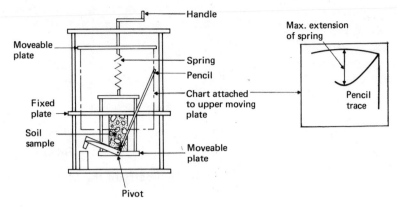

Unconfined compression test apparatus

Figure 4.19

its extension and this force is transmitted to the soil through the lower moveable plate. A chart is attached to the upper moving plate and a pencil attached to a bracket on the lower moving plate records the movement. The horizontal movement of the pencil trace is proportional to the compression of the sample and the vertical movement is equal to the extension of the spring.

Thus the maximum extension of the spring corresponds to the maximum load on the sample (P) and the principal stress

$$\sigma_1 = \frac{P(1-\varepsilon)}{a} \text{ where } \varepsilon = \text{longitudinal strain}$$

a = original cross-sectional area of sample.

Only one result is obtained from this test and since $\sigma_3 = 0$ the Mohr's circle diagram will be as shown on Fig. 4.20, and it is assumed that

$$\tau_f = c_u = \frac{\sigma_1}{2}.$$

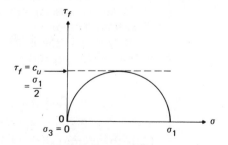

Figure 4.20

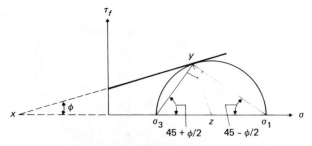

Figure 4.21

From the geometry of the Mohr's circle of stress (Fig. 4.21) it can be seen that:

$$\text{angle } xzy = 90° - \phi$$
$$\therefore \text{ angle } y\sigma_1 x = 45° - \phi/2$$
$$\therefore \text{ angle } y\sigma_3\sigma_1 = 45° + \phi/2$$

Thus if the Coulomb line is known, and the cell pressure σ_3 is given, the value of σ_1 can be found by setting out on the shear strength/pressure diagram:

$$\text{angle } y\sigma_3\sigma_1 = 45° - \phi/2$$

and angle $\sigma_3 y\sigma_1 = 90°$

This has been done on Fig. 4.18 and the following results obtained:

unconfined compression test $\sigma_3 = 0$, $\phi = 15°$

$$\sigma_1 = 150 \text{ kN/m}^2$$
$$\therefore c_u = \mathbf{75\,kN/m^2}$$

In the undrained triaxial test $\sigma_3 = 275 \text{ kN/m}^2$ $\phi = 15°$

$$\sigma_3 = 275 \text{ kN/m}^2$$
$$\therefore \sigma_1 = \mathbf{610\,kN/m^2.}$$

4.10 Change in pore-water pressure during triaxial test

In the standard triaxial test, stress changes are usually made in two stages:
(1) an all-round increase in the cell pressure ($\Delta\sigma_3$),
(2) an increase in the deviator stress ($\Delta\sigma_1 - \Delta\sigma_3$) due to an axial load increment.

Show that in the case of a saturated isotropic clay, the change in pore water pressure is given by the equation:

$$\Delta u = \frac{1}{(n \cdot C_v/C_c) + 1}[\Delta\sigma_3 + \tfrac{1}{3}(\Delta\sigma_1 - \Delta\sigma_3)]$$

where n is the porosity of the soil,
C_v is the compressibility of the pore water,
C_c is the compressibility of the soil skeleton.   (ULUC)

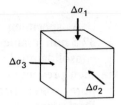

Figure 4.22

Solution Consider a cube of saturated isotropic clay of volume V (Fig. 4.22).

If it is subjected to an increase of $\Delta\sigma_1$, $\Delta\sigma_2$ and $\Delta\sigma_3$ in the three principal stresses, there will be a resulting decrease in volume $= -\Delta V$ and an increase $= \Delta u$ in the pore water pressure. The increases in the effective principal stresses will be:

$$\Delta\sigma'_1 = \Delta\sigma_1 - \Delta u$$

$$\Delta\sigma'_2 = \Delta\sigma_2 - \Delta u$$

$$\Delta\sigma'_3 = \Delta\sigma_3 - \Delta u$$

and the decrease in volume

$$-\Delta V = V \cdot \frac{1-2\mu}{E}\{\Delta\sigma'_1 + \Delta\sigma'_2 + \Delta\sigma'_3\}$$

where E = modulus of elasticity, u = Poisson's ratio.

The decrease in volume will be due to a decrease in the volume of the voids and if no drainage occurs

$$-\Delta V = n \cdot V \cdot C_v \, \Delta u \quad \text{where } n = \text{porosity of soil}$$
$$C_v = \text{compressibility of pore water.}$$

Equating the two values for $-\Delta V$

$$\therefore n \cdot V \cdot C_v \, \Delta u = V \cdot \frac{1-2\mu}{E}\{\Delta\sigma'_1 + \Delta\sigma'_2 + \Delta\sigma'_3\}.$$

In the triaxial test $\Delta\sigma_3 = \Delta\sigma_2$ and the compressibility of the soil skeleton C_c is taken as

$$C_c = \frac{3(1-2\mu)}{E} \quad \text{(assuming an elastic material)}$$

$$\therefore n \cdot C_v \cdot \Delta u = \frac{C_c}{3}\{\Delta\sigma_1 - \Delta u + \Delta\sigma_3 - \Delta u + \Delta\sigma_3 - \Delta u\}$$

$$\therefore n \cdot \frac{C_v}{C_c} \cdot \Delta u = \frac{\Delta\sigma_1}{3} + \frac{2\,\Delta\sigma_3}{3} - \Delta u$$

$$\therefore \Delta u \left(n \cdot \frac{C_v}{C_c} + 1\right) = \Delta\sigma_3 + \tfrac{1}{3}(\Delta\sigma_1 - \Delta\sigma_3)$$

$$\therefore \Delta u = \frac{1}{\left(n \cdot \dfrac{C_v}{C_c} + 1\right)}[\Delta\sigma_3 + \tfrac{1}{3}(\Delta\sigma_1 - \Delta\sigma_3)] \qquad (4.3)$$

4.11 Determination of pore-pressure parameters

A specimen of undisturbed weald clay, 75 mm long and 37·5 mm diameter, is placed in a triaxial cell and allowed to consolidate fully under a cell pressure of 200 kN/m². Drainage is then prevented and on raising the cell pressure to 400 kN/m² the pore water pressure within the sample rises to 196 kN/m². Determine the value of the pore pressure parameter B.

The following readings are then taken in a standard undrained triaxial test:

Strain dial gauge (mm)	Proving ring dial* (mm × 10⁻²)	Pore pressure (kN/m²)
0	0	196
1·5	422	290
3·0	529	290
4·5	561	276
6·0	564	263

* 1 div. = 0·5 N

Plot major and minor principal effective stresses against axial strain.

Determine the value of the pore pressure parameter A at various strains to failure.

What is the significance of the parameters A and B? (ULKC)

Solution In solution 4.10 it was shown that for a saturated isotropic clay, the change in pore water pressure resulting from a change in the principal stresses:

$$\Delta u = \frac{1}{\left(n \cdot \dfrac{C_v}{C_c} + 1\right)} [\Delta \sigma_3 + \tfrac{1}{3}(\Delta \sigma_1 - \Delta \sigma_3)] \tag{4.3}$$

This may be written in a more general form to cover other soil states:

$$\Delta u = B[\Delta \sigma_3 + A(\Delta \sigma_1 - \Delta \sigma_3)] \tag{4.4}$$

where A and B are referred to as the *pore pressure parameters*. If the soil is subjected to an all-round pressure increase $\Delta \sigma_3$,

$$\Delta u = B \, \Delta \sigma_3$$

$$\therefore \ B = \frac{\Delta u}{\Delta \sigma_3}.$$

For a saturated soil since the water is relatively incompressible $B = 1$. For a dry soil since the air in the voids is highly compressible $B = 0$ and for a partially saturated soil B will have a value between 0 and 1.

It can be seen from eqn (4.3) that:

$$B = \frac{1}{\left(n \cdot \dfrac{C_v}{C_c} + 1\right)}$$

and is therefore dependent on the compressibility of the soil structure relative to that of the water in the voids.

If the soil is then subjected to an increment of pressure in the vertical direction only, for a perfectly elastic solid $A = \frac{1}{3}$. In fact A varies considerably since the soil is not elastic and is normally determined experimentally. A may have values varying from more than $+1$ to $-0\cdot5$.

From the given test results:
During the initial consolidation process

$$\Delta\sigma_3 = \Delta\sigma_1 = 200 \text{ kN/m}^2$$

$$\Delta u = u - u_p \text{ where } u_p = \text{initial pore water pressure}$$

$$= 0$$

$$\therefore \Delta u = 196 \text{ kN/m}^2$$

$$\Delta u = B[\Delta\sigma_3 + A(\Delta\sigma_1 - \Delta\sigma_3)]$$

$$\therefore 196 = B[200 + A(0)]$$

$$\therefore B = \mathbf{0\cdot98.}$$

For the remainder of the test no drainage is allowed.

Axial strain $\varepsilon = \dfrac{\Delta l}{l}$

Axial stress $(\sigma_1 - \sigma_3) = \text{load on proving ring} \times \dfrac{(1 - \varepsilon)}{A}$

$$\sigma_1 = (\sigma_1 - \sigma_3) + \sigma_3$$

$$\sigma_1' = \sigma_1 - u$$

$$\sigma_3 = \text{cell pressure}$$

$$\sigma_3' = \sigma_3 - u.$$

Using these relationships, the following values are calculated:

Δl (mm)	ε	$\sigma_1 - \sigma_3$ (kN/m²)	σ_1 (kN/m²)	σ_3 (kN/m²)	u (kN/m²)	σ_1' (kN/m²)	σ_3' (kN/m²)
0	0	0	400	400	196	204	204
1·5	0·02	229	629	400	290	339	110
3·0	0·04	279	679	400	290	389	110
4·5	0·06	291	691	400	276	415	124
6·0	0·08	286	686	400	263	423	137

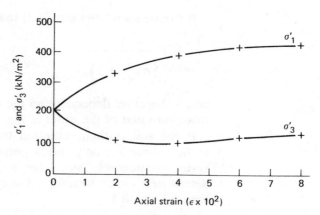

Figure 4.23

Using the appropriate values from this table, the major principal effective stress σ_1 and the minor principal effective stress σ_3 are plotted against axial strain on Fig. 4.23.

The pore pressure parameter A can be calculated from the measured values of the changes in σ_1, σ_3 and u using eqn (4.4).

Pressure range (σ_1)	$\Delta\sigma_1$	$\Delta\sigma_3$	Δu	B	A
400–629	229	0	94	0·98	0·42
400–679	279	0	94	0·98	0·34
400–691	291	0	70	0·98	0·25
400–686	286	0	67	0·98	0·19

The significance of A and B was explained at the start of the solution. The pore water parameters are used to calculate the changes in pore water pressure and hence effective pressures which occur during the construction of earthworks. These stresses are used in stability calculations to find the factor of safety.

4.12 Cohesion of a soil sample using the unconfined compression test

A sample of clay is tested in an autographic recording unconfined compression testing machine, and the trace obtained is shown on Fig. 4.24. Determine the value of C_u given the following particulars:

Specimen: original length 75 mm
original diameter 37·5 mm
Machine: length of recording arm from pivot to pencil point 225 mm
length of recording arm from pivot to base 62·5 mm
spring stiffness 8·75 N/mm

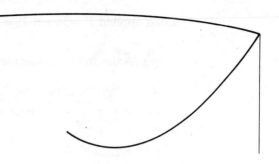

Figure 4.24

Solution The figure shows a typical result obtained from an unconfined compression test. The horizontal distance across the chart records the compression of the specimen, to a magnified scale. In this case the recording arm ratio is $225:62\cdot5 = 3\cdot6:1$.

∴ 1 mm along an arc line corresponds to a shortening of the specimen of $1/3\cdot6$ mm.

The vertical movement records the extension of the spring.

∴ Load applied = vertical movement × stiffness of spring.

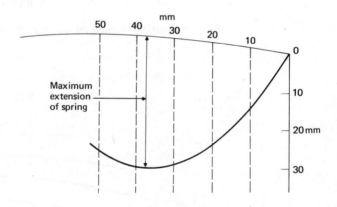

Figure 4.25

Scaling off Fig. 4.25, the maximum extension of the spring = 34 mm at a distance 38 mm in from 0.

$$\therefore \text{ Compression of sample} = \frac{38}{3\cdot6} = 10\cdot55 \text{ mm}$$

$$\therefore \text{ Longitudinal strain } \varepsilon = \frac{10\cdot55}{75} = 0\cdot141.$$

Load applied $P = 34 \times 8.75 = 298 \, \text{N}$

$\therefore$ Axial stress on sample $\sigma_1 = \dfrac{P(1-\varepsilon)}{a}$

$$= \frac{298(1-0.141)}{\dfrac{\pi}{4} \times 37.5^2} \frac{\text{N}}{\text{mm}^2} \left[\frac{\text{kN}}{10^3 \, \text{N}} \times \frac{10^6 \, \text{mm}^2}{\text{m}^2} \right]$$

$$= 232 \, \text{kN/m}^2.$$

In an unconfined compression test on a saturated clay, the shearing resistance is taken as half the applied axial stress at failure (Fig. 4.20).

$$\therefore \ \tau_f = C_u = \frac{232}{2} = 126 \, \text{kN/m}^2.$$

It is usual to have a transparent mask which can be placed over the pencil trace and from which the axial stress can be calculated.

The stress on the sample $\sigma_1 = \dfrac{\text{extension of spring} \times \text{stiffness } \lambda}{\text{area of sample } a}$

$$\therefore \ \text{Extension} = a \cdot \frac{\sigma_1}{\lambda} = 1\,100 \frac{\sigma_1}{\lambda} \ \text{for a standard 62.5 mm dia.}$$
sample.

By substituting values of σ_1, say 5, 10, 15, 20, 25 mm etc., a series of

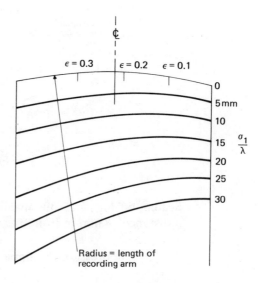

Figure 4.26

extensions are obtained. These represent the vertical distances measured from the origin on the mask for no deformation of the specimen, i.e. down the right hand side of the mask.

The mask would then consist of a series of circular arcs drawn from these points if the specimen did not deform. To allow for the deformation of the specimen:

$$\text{Extension} = \frac{1\,100}{(1-\varepsilon)} \cdot \frac{\sigma_1}{\lambda} \quad \text{where } \varepsilon = \text{axial strain}$$

By substituting the same values of $\frac{\sigma_1}{\lambda}$ for various values of ε a series of points are obtained which represent the vertical distances on the mask below the horizontal distance corresponding to ε. The mask is completed by joining equal values of $\frac{\sigma_1}{\lambda}$.

A typical mask is shown on Fig. 4.26.

4.13 Shearing resistance of a soil mass using a vane shear test

> Describe briefly the vane shear test.
>
> Assuming that the soil shears along the surface of the cylinder of revolution of the vane and that the distribution of shear stress is uniform across the ends of the cylinder, show that the torque required to rotate a vane of diameter d and height h in a soil of cohesive strength c is given by
>
> $$T = c\pi \left(\frac{d^2 h}{2} + \frac{d^3}{6} \right)$$
>
> A vane is 100 mm diameter and has blades 150 mm long. When pushed into undisturbed soil at the bottom of a bore hole, the torque required to rotate the vane is 190 Nm. What is the shearing strength of the soil? (HNC)

Solution One difficulty encountered when finding the shearing resistance of some soils is the problem of obtaining undisturbed samples. A way in which this may be overcome is by means of the vane test. The vane consists essentially of four steel plates welded at right angles and attached to a steel torque rod. The vane is forced into the soil and a torque is applied until the soil fails. The torque is measured by measuring the angle of twist. Although essentially an *in-situ* test, a laboratory vane of smaller dimensions is also used. A diagram of the arrangement is shown on Fig. 4.27(*a*).

The assumed shear stress distribution round a vane blade is shown on Fig. 4.27(*b*). Thus the shearing resistance of the soil along the sides of the vane is $c \cdot \pi \cdot d \cdot h$, and at each end $c \cdot \dfrac{\pi d^2}{4}$.

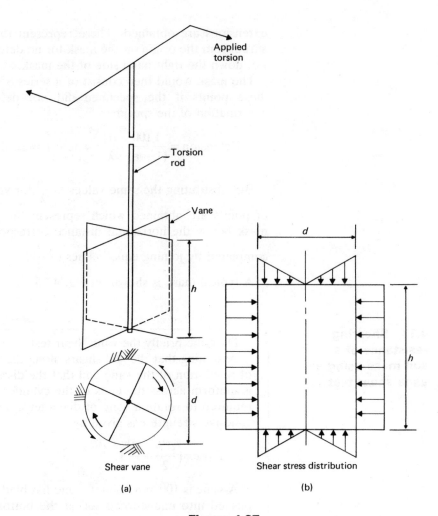

Applied torsion

Torsion rod

Vane

h

d

Shear vane

d

Shear stress distribution

(a)

(b)

Figure 4.27

∴ The torsional resistance about the centre

$$T = c \cdot \pi \cdot dh \times \frac{d}{2} + \left(c \cdot \frac{\pi}{4} d^2 \times \frac{2}{3} \frac{d}{2} \right)^2$$

$$= c\pi \left(\frac{d^2 h}{2} + \frac{d^3}{6} \right).$$

For the given vane $d = 100 \text{ mm} = 1 \times 10^{-1} \text{ m}$

$$h = 150 \text{ mm} = 1 \cdot 5 \times 10^{-1} \text{ m}$$

$$T = 190 \text{ Nm} = 1 \cdot 90 \times 10^{-1} \text{ kNm}$$

$$\therefore 1 \cdot 90 \times 10^{-1} = c\pi \left\{ \frac{(1 \times 10^{-1})^2 (1 \cdot 5 \times 10^{-1})}{2} + \frac{(1 \cdot 5 \times 10^{-1})^3}{6} \right\}$$

$$\therefore c = 46 \text{ kN/m}^2.$$

4.14 Comparison of shearing resistance of a soil mass using different size vanes

Two vanes A and B with length to diameter ratios of 3 and 1 respectively, are each used to make a test at the same depth in a clay layer. Both vanes have a diameter of 50 mm and the torques measured were 955 kN mm for A and 445 kN mm for B. Find the values of apparent cohesion on horizontal and vertical planes at this depth.

Explain why the apparent cohesion may vary with direction.

(CEI)

Solution The assumed shear stress distribution round a vane blade is shown on Fig. 4.27(b).

Let c_v = apparent cohesion on vertical plane
$\quad c_h$ = apparent cohesion on horizontal plane

For vane A, $h = 150$ mm, $d = 50$ mm, $T = 955$ kN mm

$$\therefore T = c_v \cdot dh \times \frac{h}{2} + 2\left(c_h \cdot \frac{\pi}{4} \cdot d^2 \times \frac{2}{3} \cdot \frac{d}{2}\right)$$

$$\therefore 955 = c_v \cdot \pi \cdot 50 \cdot \frac{150^2}{2} + 2\left(c_h \cdot \frac{\pi}{4} \times \frac{50^3}{3}\right)$$

$$\therefore \frac{955}{50^3 \times \pi} = \frac{9c_v}{2} + \frac{c_h}{6} \qquad (a)$$

For vane B, $h = 50$ mm, $d = 50$ mm, $T = 445$ kN mm

$$\therefore 445 = c_v \cdot \pi \frac{50^3}{2} + 2\left(c_h \cdot \frac{\pi}{4} \times \frac{50^3}{3}\right)$$

$$\therefore \frac{445}{50^3 \times \pi} = \frac{c_v}{2} + \frac{c_h}{6} \qquad (b)$$

$(a) - (b)$: $\dfrac{955 - 445}{50^3 \times \pi} = 4c_v$

$$\therefore c_v = \frac{510}{50^3 \times \pi \times 4} = 0 \cdot 325 \times 10^{-3} \text{ kN/mm}^2 = \textbf{325 kN/m}^2.$$

Substituting in (b)

$$\therefore c_h = \textbf{5 823 kN/m}^2.$$

The apparent cohesion may vary with direction as a result of the formation of the clay which may be laminated.

4.15 Elastic modulus and Poisson's ratio of a soil sample

Data from an undrained triaxial test on an undisturbed sample are as follows:

σ_1 (kN/m²)	35·0	51·8	70·0	99·4	124·6	128·8
σ_3 (kN/m²)	35·0	35·0	35·0	35·0	35·0	35·0
ε (%)	0	0·25	0·5	1·0	2·0	3·0

This sample has been taken from a thick layer of clay on which a footing is to be placed. The immediate settlement is given by the expression $\dfrac{qB(1 - \mu^2)I_p}{E}$ where q, B and I_p define the pressure and shape of the foundation, E is the elastic modulus and μ is the Poisson's ratio of the soil. Determine the appropriate values of E and μ, making appropriate assumptions.

Explain how the estimates of immediate settlement may be used in conjunction with the results of tests in a consolidation press to give a better estimate of the total settlement at any time after the loading of the foundations. Include a sketch of settlement against time in your answer. Ignore non-elastic lateral strains and omit any description of how the average degree of consolidation U is obtained. (CEI)

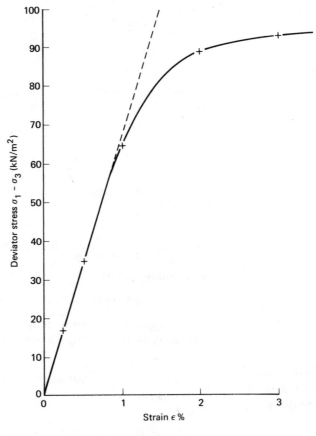

Figure 4.28

Solution The average of E may be estimated by plotting the deviator stress $(\sigma_1 - \sigma_3)$ against strain (Fig. 4.28).

The first portion of this graph is practically a straight line and may therefore be used in the calculation $E = \dfrac{\text{stress}}{\text{strain}}$

$$E = \frac{100}{1\cdot45 \times 10^{-2}} = 6\,900\ \text{kN/m}^2.$$

(If there is no straight portion to the curve, a straight line may be drawn from the origin to the point on the curve representing the stress expected to be imposed by the foundation and the 'secant modulus' calculated). For a saturated soil Poisson's ratio μ is normally assumed to be 0·5.

The total settlement of a foundation may be considered to be made up of

$$\rho_v = \rho_i + \rho_c$$

where ρ_i = immediate settlement calculated from formula given

ρ_c = consolidation settlement calculated from oedometer tests (Chapter 3)

Figure 4.29 shows a typical settlement/time curve. Initially, there may be some swelling of the soil due to relief of pressure during excavation.

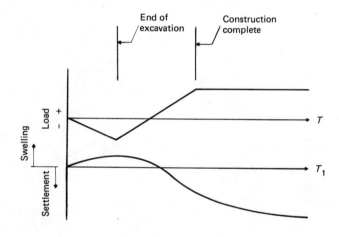

Figure 4.29

4.16 Shear stress in a soil mass under a flexible foundation

It is proposed to construct at ground level a long flexible tank 18 m wide, which will impose a uniform loading of 46 kN/m². The site is underlain by a considerable thickness of normally consolidated silts and clays, of which the top 3·5 m has weathered and dried. The mean saturated density of the soil is 1·84 Mg/m³ and the water table stands at 1·25 m below ground level.

The shear strength c_u of the soil below the crust can be expressed as $c_u/p = 0.2$ where p is the effective overburden pressure before loading. In the crust, the strength may be assumed to be adequate to resist the shear stresses imposed by the tank.

The maximum shear stress at a point X below a strip loading of intensity q is given by $(q/\pi) \sin \beta$, where β is defined as in Fig. 4.30.

Plot the distribution of maximum shear stress under the tank (to the left of the centre line only) using $\beta = 30$, 50, 70, 90, 110 and 130°. Plot the zone in which the soil is overstressed in shear.

Discuss the feasibility of the project.   (ICE)

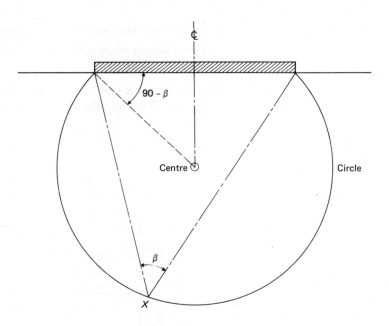

Figure 4.30

Solution The expression given for the maximum shear stress

$$\tau = \frac{q}{\pi} \cdot \sin \beta$$

is a development from Boussinesq's formula for the vertical normal stress [eqn (3.1)] and is based on the same assumptions i.e. the soil is elastic, homogeneous, isotropic and semi-infinite in extent, together with the further assumption that the loading is a long narrow strip.

Substituting in the formula for the different values gives:

β (°)	τ (kN/m²)
30	7·32
50	11·21
70	13·76
90	14·64
110	13·76
130	11·21

The circles of stress for these values can then be drawn using the construction indicated on Fig. 4.31.

The actual shear strength of the soil $c_u = 0·2p$ where p is the effective overburden pressure due to the weight of overlying soil. In order to find if the soil is overstressed, it is convenient to use the values of c_u corresponding to the values of τ and calculate the depths z below the ground surface at which they occur.

Hence, below the water table:

$$p = \gamma_{sat} \times 1·25 + (\gamma_{sat} - \gamma_w)(z - 1·25)$$
$$= (\gamma_{sat} - \gamma_w)z + 1·25\gamma_w$$

and since $c_u = 0·2p$

$$\therefore \frac{c_u}{0·2} = (\gamma_{sat} - \gamma_w)z + 1·25\gamma_w$$

$$\therefore z = \frac{\dfrac{c_u}{0·2} - 1·25\gamma_w}{(\gamma_{sat} - \gamma_w)}$$

and

$$\gamma_{sat} = 1·84 \times 10 = 18·4 \text{ kN/m}^2, \qquad \gamma_w = 10 \text{ kN/m}^3$$

$$\therefore z = \frac{\dfrac{c_u}{0·2} - 1·25 \times 10}{(18·4 - 10·0)} = 0·595c_u - 1·488$$

Substituting in this formula gives:

c_u (kN/m²)	z (m)
7·32	2·87
11·21	5·18
13·76	6·70
14·64	7·22

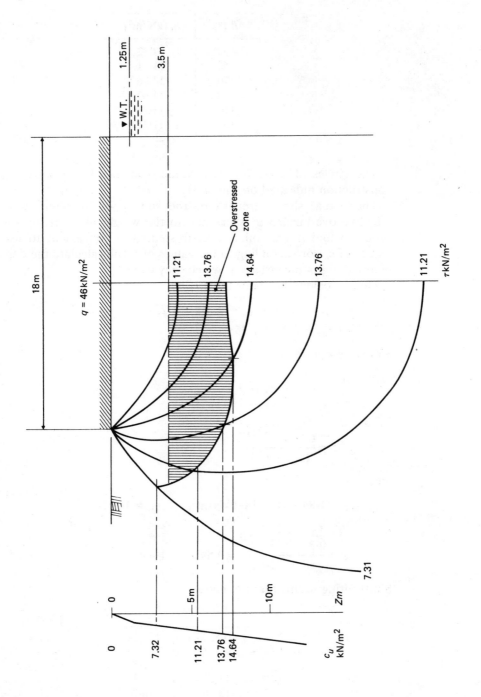

Figure 4.31

If these values are plotted to the same scale on the stress circle diagram, the points where the shear stress is equal to the shear strength can be found in the manner indicated on Fig. 4.31.

If it is assumed that the crust is sufficiently strong to a depth of 3·5 m as the question indicates, the overstressed zone is as shown with vertical shading on Fig. 4.31.

Problems

1. The following results were obtained from undrained shear box tests on specimens of sandy clay of cross-section 60 mm by 60 mm square.

Normal load (kN)	Shearing force at failure (kN)
1·00	0·47
0·50	0·32
0·25	0·24

Find the apparent cohesion and angle of shearing resistance.

How can the results of such tests be applied in determining the safe bearing capacity for a foundation on this soil?

If a triaxial test is carried out on a specimen of the same soil with a cell pressure of $140 \, kN/m^2$, find the total axial stress at which failure would be expected. (ICE)

($46 \, kN/m^2$, 17°, $380 \, kN/m^2$)

2. Sketch typical results to be expected in the following triaxial tests:
  (a) Undrained test on a saturated clay.
  (b) Undrained test on a sample of embankment fill.
  (c) Consolidated-undrained test on an overconsolidated clay.
  (d) Drained test on a dense sand.

Indicate in each case a practical problem to which the results of the test could be applied.

What are the limitations of the shear box test? In such a test on a dense dry sand the following results were obtained:

Vertical stress	kN/m^2	155;	292;	430.
Maximum shear stress	kN/m^2	117;	200;	296.

Determine the shear strength parameters. For the second test determine the magnitude and direction of the principal stresses in the specimen at failure. Comment on the volume changes to be expected during these tests. (ICE)

($0 \, kN/m^2$, 35°; $676 \, kN/m^2$ on plane 62° to horizontal, $188 \, kN/m^2$ on plane 28° to horizontal)

3. Define the angle of shearing resistance of a soil and explain why this is not always the same as the angle of internal friction.

The following results were obtained from undrained triaxial tests on three specimens from an undisturbed sample of soil.

Lateral pressure (kN/m²)	70	140	210
Total vertical stress at failure (kN/m²)	224	294	371

Plot the Mohr circles representing these results and determine the apparent cohesion and the angle of shearing resistance.

What type of soil would you expect to give results such as these? (ICE)

(77 kN/m^2; $1\cdot5°$; clay)

4. Define effective stress and comment on its importance in practical soil mechanics problems.

From triaxial tests with pore water measurement it is found that the apparent cohesion and angle of shearing resistance of a soil referred to effective stress, are 70 kN/m^2 and $25°$ respectively. Find the shearing strength of this soil at a depth of 9 m below ground surface; the soil is fully saturated (average density being $1\cdot92 \text{ Mg/m}^3$) and the water table is at a depth of $2\cdot75$ below the surface. (ICE)

(125 kN/m^2)

5. The following readings were taken during a shear box test on a sample of dense sand, the box being 60 mm^2:

Time (min)	0	2	4	6	8	10	12
Proving ring dial* (10^{-3} mm)	0	142	218	247	263	254	249

* 1 div. = $1\cdot5$ N.

The vertical load was 450 N and the screw moved the lower box at $0\cdot3$ mm/min. Plot shearing stress against the true displacement and determine the value of the angle of shearing resistance. By means of a Mohr diagram determine the magnitude and direction of the principal stresses at failure. (HNC)

($39°$; $\sigma_1 = 334 \text{ kN/m}^2$ at $66°$ to horizontal, $\sigma_3 = 78 \text{ kN/m}^2$ at $34°$ to horizontal)

6. Describe, with neat sketches, a laboratory apparatus for pore water pressure measurement in soils.

The following readings were taken in an undrained triaxial test on a sample of saturated sand originally 75 mm long and 37·5 mm diameter, consolidated to 35 kN/m^2:

Change in length (1/40 mm)	Proving ring dial gauge*	Pore water pressure (kN/m^2)	Cell pressure (kN/m^2)
		0	35
0	0	490	525
250	157	434	
500	280	133	
750	355	35	
1 000	402	14	

* 1 div. = 5 N.

Plot $(\sigma_1' - \sigma_3')$ and (σ_1'/σ_3') against axial strain, and calculate values of ϕ' at failure from the peak of each curve.

Why is the cell pressure increased at the start of the test?

(ULKC)

$(34°, 35°)$

7. A deep deposit of fine sand has a mean void ratio of 0·650 and is composed of material having a grain specific gravity of 2·65. Compute the unit weight of the sand when dry, when 50% saturated and when fully saturated.

Draw a diagram showing the distribution of effective vertical pressure on horizontal sections through the deposit to a depth of 6 m, if the water table stands at 1·25 m below ground level. The soil is fully saturated above the water table.

Due to drainage, the water table is lowered to 3 m below ground level. Estimate the percentage increase in shear strength of the sand at a depth of 2·5 m, assuming that the soil remains fully saturated for a height of 1·5 m above the water table, and that above this level it has a mean degree of saturation of 50%. Take $c' = 0$. (ICE)

$(\gamma_d = 16·1 \text{ kN/m}^3, \ \gamma_{sat} = 20·0 \text{ kN/m}^3, 25\%)$

8. Describe the essential features of the triaxial test. Indicate how to carry out a consolidated undrained test, including pore water pressure measurement in the second stage of the test.

For a certain saturated normally consolidated soil stratum the coefficient of earth pressure at rest (defined as the ratio of *effective* horizontal and vertical stresses) is 0·6, and the shear strength parameters are $c' = 0$, $\phi' = 20°$. Also it has been established in general that the change in pore water pressure u caused by the changes $\Delta\sigma_1$ and $\Delta\sigma_3$ occurring in the horizontal and vertical total stresses acting on a material is given by $\Delta u = \Delta\sigma_3 + A(\Delta\sigma_1 - \Delta\sigma_3)$.

If the value of A is 0·5 during sampling operation, determine the effective stresses in a laboratory specimen of the soil obtained from a sample removed from a location in the stratum where the effective overburden pressure is $110 \, \text{kN/m}^2$.

In an unconfined compression test on the sample the axial stress at failure was $50 \, \text{kN/m}^2$. Determine the change in the pore pressure during shear. (ICE)

$(88 \, \text{kN/m}^2; \, 40 \, \text{kN/m}^2)$

9. Write down Skempton's equation for the change in pore pressure which occurs when a soil is subjected to changes in the total principal stresses σ_1 and σ_3. What is the significance of the equation? Specimens of a certain saturated clay soil were consolidated in the triaxial machine under all-round pressures σ_3 and were then sheared without further drainage, at axial stresses σ_1. The following results were obtained:

All round pressure during consolidation and shear (kN/m^2)	Value of pore pressure parameter A at failure
28	−0·19
360	0·44

The soil is known to have the properties $c' = 28 \, \text{kN/m}^2$ and $\cdot\phi' = 23°$.

Determine the shear stress parameters in terms of total stress. Would you classify the soil as overconsolidated or normally consolidated? (ICE)

$(12° \, 57', \, 56 \, \text{kN/m}^2)$

10. At a certain location borings indicate a 3-m layer of coarse sand (having a mean void ratio of 0·63 and grain specific gravity 2·70) overlying a thick layer of normally consolidated clay with a mean bulk density of $2·0 \, \text{Mg/m}^3$. Above the water table stands at 1·5 m below ground level, the sand is dry.

The data obtained in a consolidated-undrained test on clay samples taken from 9 m below ground level are:

Cell pressure during consolidation (kN/m^2)	At failure	
	Principal stress difference (kN/m^2)	Pore water pressure (kN/m^2)
462	269	255
689	400	379

Determine the shear strength parameters in terms of effective stress, and the value of the pore pressure parameter A.

Estimate the shear strength likely to be obtained from an unconfined compression test on another sample from the same level. (ICE)

$(c' = 0,\ \phi' = 22.5°,\ A = 0.95,\ c_u = 28\,\text{kN/m}^2)$

5

Lateral soil pressures

The pressure at any point in a fluid such as water is the same in all directions. Thus the lateral pressure on a vertical surface retaining water is equal to $\gamma_w h$ where $h =$ the head of water above the point considered. Figure 5.1 shows the lateral pressure diagram on a wall of height H retaining water.

The total force P per unit length of wall will be equal to the area of the pressure diagram.

$$\therefore \ P = \tfrac{1}{2}\gamma_w \cdot H^2$$

and this force will act at the centroid of the diagram i.e. $\tfrac{2}{3}H$ from the surface.

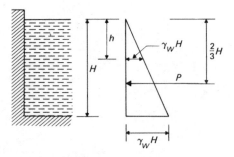

Figure 5.1

In the case of soil, which unlike water, possesses resistance to shearing, the lateral pressure at any point will not be the same as the vertical pressure at that point. The lateral pressure will vary depending on whether the soil is free to move laterally or is subjected to a lateral compressive force.

In order to design soil retaining structures such as retaining walls and sheet pile walls, it is necessary to determine the magnitude of the lateral pressures to which the structure is subjected.

Worked examples

5.1 Active and passive soil pressures on a vertical surface

Explain what is meant by the term 'state of plastic equilibrium of a soil mass', and hence distinguish between active and passive earth pressures. How must a retaining wall yield so that the active earth pressure of the backfill is properly mobilized?

A vertical wall of height H retains a backfill having a horizontal surface level with top of the wall. The backfill is an intact clay having a unit weight γ and the shear strength parameters $c_u = c$, $\phi_u = 0$. The adhesion between the wall and the clay is expressed as mc, where m is a coefficient.

Show that in the active state before changes in water content have occurred the maximum horizontal pressure sustained by the wall is:

$$\frac{\gamma \cdot H^2}{2} - 2Hc(1+m)^{1/2}$$

Make an estimate of the initial maximum safe height of an unsupported vertical cut in this clay. (ICE)

Solution A state of plastic equilibrium exists in a soil mass when every part of it is just on the point of failing in shear, i.e. the particles are on the point of moving relative to one another.

In its natural state, an element of soil at a depth z will be subjected to a vertical pressure $\sigma_v = \gamma z$ and a corresponding horizontal pressure $\sigma_{h0} = K_0 \sigma_v$ where K_0 is the coefficient of earth pressure at rest and is equal to the ratio of $\dfrac{\sigma_{h0}}{\sigma_v}$.

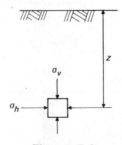

Figure 5.2

If the soil is free to move laterally, the horizontal pressure will decrease as the shearing resistance of the soil is mobilized until a state of plastic equilibrium is reached. The horizontal pressure will then be $\sigma_{ha} = K_a \sigma_v$ where K_a is the coefficient of *active* earth pressure. σ_{ha} is called the active earth pressure since it is the lateral pressure exerted by the soil when all the shearing resistance has just been fully mobilized. If the soil is allowed to yield still further, there will be no change in σ_{ha} but sliding will occur between the soil particles.

If the soil is subjected to lateral compression, the horizontal resistance will increase as the shearing resistance is mobilized until a state of plastic equilibrium is reached when $\sigma_{hp} = K_p\sigma_v$, where K_p is the coefficient of *passive* earth resistance. σ_{hp} is called the passive earth resistance (or pressure) and represents the maximum resistance to lateral displacement which the soil can withstand when the shearing resistance is fully mobilized. Further compression will not increase σ_{hp} but will lead to movement between the soil particles.

These two states represent the limiting conditions of the soil for equilibrium. Between these two limits the soil is in a state of *elastic* equilibrium.

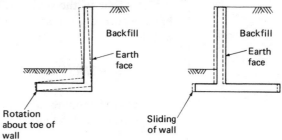

Figure 5.3

In order to mobilize the active pressure of the backfill σ_{ha}, the earth face of a retaining wall must yield as shown in dotted lines on Fig. 5.3.

The assumption is made that as the wall yields away from the backfill, the soil is brought to the point of active plastic equilibrium. In this condition a wedge of soil is about to break away from the main soil mass as shown on Fig. 5.4(a).

To prevent this happening the load from the wedge (W) must be

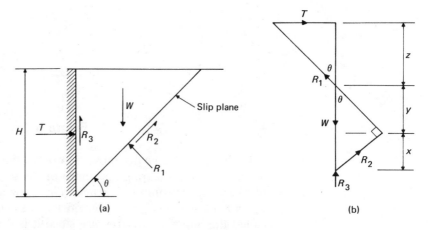

Figure 5.4

kept in equilibrium by the forces acting on it. These are:

T = thrust from retaining wall (acting at right angles to wall if $\phi = 0$)

R_1 = reaction between the wedge and the main soil mass (acting at right angles to the slip plane if $\phi = 0$)

R_2 = cohesion along the slip plane = $c \times$ length of slip plane

R_3 = cohesion between the wall and the soil = $mc \times$ height of wall in contact with soil.

For any particular wedge, W, R_2 and R_3 can be calculated and the line of action of all the forces is known. Thus by drawing W, R_2 and R_3 to scale and from the ends of the resulting diagram, drawing lines parallel to the lines of action of R_1 and T a polygon of forces will be completed from which the value of T can be scaled [Fig. 5.4(b)].

From the figure the following relationships can be found:

$$W = \tfrac{1}{2} . H . \frac{H}{\tan \theta} . \gamma = \frac{H^2 \gamma}{2 \tan \theta}$$

$$R_2 = c . \frac{H}{\sin \theta}$$

$$R_3 = mc . H$$

Then from these values and knowing the line of action of T and R_1 the polygon of forces is drawn.

From Fig. 5.4(b):

$$x = R_2 \sin \theta = cH$$

$$y = \frac{R_2 \cos \theta}{\tan \theta} = \frac{cH}{\tan^2 \theta}$$

$$z = \frac{T}{\tan \theta}$$

$$\therefore x + y + z = cH + \frac{cH}{\tan^2 \theta} + \frac{T}{\tan \theta} = W - R_3$$

$$= \frac{H^2 \gamma}{2 \tan \theta} - mc . H$$

$$\therefore T = \frac{H^2 \gamma}{2} - mc . H . \tan \theta - cH \tan \theta - \frac{cH}{\tan \theta}$$

For max. value of T: $dT/d\theta = 0$

$$\therefore 0 = -mcH \sec^2 \theta - cH \sec^2 \theta + cH \frac{\sec^2 \theta}{\tan^2 \theta}$$

$$\therefore \tan \theta = \frac{1}{(m + 1)^{1/2}}$$

$$\therefore T_{max} = \frac{H^2\gamma}{2} - \frac{mcH}{(m+1)^{1/2}} - \frac{cH}{(m+1)^{1/2}} - cH(m+1)^{1/2}$$

$$= \frac{H^2\gamma}{2} - cH\frac{m+1+m+1}{(m+1)^{1/2}}$$

$$= \frac{H^2\gamma}{2} - 2cH(m+1)^{1/2}$$

An estimate of the safe height to which an unsupported vertical cut could initially be made is found using the above formula and putting $T = 0$. Then with no wall present $m = 0$

$$\therefore 0 = \frac{H^2\gamma}{2} - 2cH$$

$$\therefore \text{ Height of cut } H = \frac{4c}{\gamma}. \qquad (5.1)$$

5.2 Active and passive soil pressures using the Mohr's circle

> With the aid of Mohr's circle diagrams explain what is meant by active and passive Rankine states in a '$c - \phi$' soil with a horizontal surface. Hence obtain an expression for the intensity of active pressure exerted by such a soil at a depth behind a retaining wall with smooth, vertical back. The surface of the soil is horizontal and coincides with the top of the wall. (ICE)

Solution An element of soil at a depth z below a horizontal ground surface (Fig. 5.2) is subjected to a vertical stress $\sigma_v = \gamma z$ and a horizontal stress σ_h. Since the element is symmetrical with reference to a vertical plane both these stresses will be principal stresses and can be represented on a Mohr's circle diagram.

If the soil is in a state of active plastic equilibrium (described in solution 5.1), it is just on the point of failing in shear and consequently the Mohr's circle for this state will just be tangential to the line representing the Coulomb shear strength equation for the soil (Fig. 5.5). This condition is sometimes referred to as the active Rankine state.

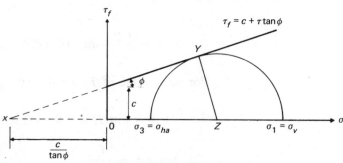

Figure 5.5

Then $\sigma_h = \sigma_{ha}$ the active pressure of the soil. It was established from the geometry of the diagram that

$$\sigma_1 = \sigma_3\left(\frac{1+\sin\phi}{1-\sin\phi}\right) + \frac{2c\cos\phi}{1-\sin\phi} \qquad \text{[eqn (4.2)]}$$

In this case

$$\sigma_1 = \sigma_v = \gamma z, \qquad \sigma_3 = \sigma_{ha}$$

$$\therefore \gamma z = \sigma_{ha}\left(\frac{1+\sin\phi}{1-\sin\phi}\right) + \frac{2c\cos\phi}{1-\sin\phi}$$

$$\therefore \sigma_{ha} = \gamma z\left(\frac{1-\sin\phi}{1+\sin\phi}\right) - \frac{2c\cos\phi}{1+\sin\phi}$$

$$\therefore \sigma_{ha} = \gamma z\left(\frac{1-\sin\phi}{1+\sin\phi}\right) - 2c\sqrt{\frac{1-\sin^2\phi}{1+\sin\phi}}$$

$$\therefore \sigma_{ha} = \gamma z\left(\frac{1-\sin\phi}{1+\sin\phi}\right) - 2c\sqrt{\frac{1-\sin\phi}{1+\sin\phi}} \qquad (5.2)$$

where σ_{ha} = active pressure exerted by a '$c-\phi$' soil i.e. one having apparent cohesion c and an angle of shearing resistance ϕ.

If the soil is in a state of passive plastic equilibrium, the corresponding Mohr's circle diagram will be as shown on Fig. 5.6 and $\sigma_h = \sigma_{hp}$ and is the major principal stress. This will be the passive Rankine state.

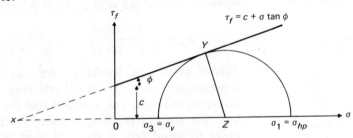

Figure 5.6

In this case

$$\sigma_1 = \sigma_{hp} \text{ the passive resistance of a '} c-\phi \text{' soil}$$

$$\sigma_3 = \sigma_v = \gamma z$$

and from the geometry of the figure:

$$\sigma_{hp} = \gamma z\left(\frac{1+\sin\phi}{1-\sin\phi}\right) + \frac{2c\cos\phi}{1-\sin\phi}$$

$$\therefore \sigma_{hp} = \gamma z\left(\frac{1+\sin\phi}{1-\sin\phi}\right) + \frac{2c\sqrt{1-\sin^2\phi}}{1-\sin\phi}$$

$$\therefore \sigma_{hp} = \gamma z\left(\frac{1+\sin\phi}{1-\sin\phi}\right) + 2c\sqrt{\frac{1+\sin\phi}{1-\sin\phi}} \qquad (5.3)$$

Equations (5.2) and (5.3) are known as Bell's equations for the active and passive pressures of a $c - \phi$ soil.

For a cohesionless soil ($c = 0$) the corresponding values become

$$\sigma_{ha} = \gamma z \left(\frac{1 - \sin \phi}{1 + \sin \phi} \right)$$

$$\sigma_{hp} = \gamma z \left(\frac{1 + \sin \phi}{1 - \sin \phi} \right)$$

These are generally referred to as Rankine's equations.

5.3 Active pressure of a cohesionless soil on a vertical wall

A retaining wall 6 m high, with earth face vertical, supports cohesionless soil of dry density 1·6 Mg/m³, angle of shearing resistance 35° and void ratio 0·68. The surface of the soil is horizontal and level with the top of the wall. Neglecting wall friction, determine the total earth thrust on the wall/lineal foot: (a) if the soil is dry; (b) if owing to inadequate drainage it is waterlogged to a level 2·5 m below the surface.

Find also at what height above the base of the wall the thrust acts. (ICE)

Solution　From eqn (5.2), since this is a cohesionless soil $c = 0$, and friction between the wall and soil is neglected:

$$\therefore \ \sigma_{ha} = \gamma z \frac{1 - \sin \phi}{1 + \sin \phi}$$

This is known as Rankine's formula for the active pressure of a cohesionless soil on a vertical retaining wall. In working out the earth thrust on the wall, it is assumed that the wall is free to move in such a way that the soil reaches its state of active plastic equilibrium and in order to avoid failure the wall must be capable of withstanding the thrust which the soil is exerting.

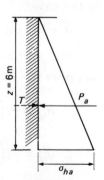

Pressure distribution diagram

Figure 5.7

(*a*) Consider 1 m run of wall.

$$\rho_d = 1 \cdot 6 \, \text{Mg/m}^3 \quad \therefore \quad \gamma_d = 16 \cdot 0 \, \text{kN/m}^2, * \quad z = 6 \, \text{m}$$

Then $\sigma_{ha} = 16 \times 6 \times \dfrac{1 - \sin 35°}{1 + \sin 35°}$

$$= 26 \, \text{kN/m}^2$$

Total thrust P_a = area of pressure distribution diagram (Fig. 5.7)

$$= \frac{26 \times 6}{2} = \textbf{78 kN/ m}^2 \textbf{ run of wall.}$$

It will act at the centroid of the diagram

i.e. $\dfrac{6}{3}$ m = **2 m up from base.**

(*b*) If the soil is waterlogged to a level 2·5 m below the top of the wall, the active pressure due to the soil below the water table will be related to the submerged density ρ' instead of the dry density but in addition there will be the hydrostatic pressure due to the water. This is shown on Fig. 5.8.

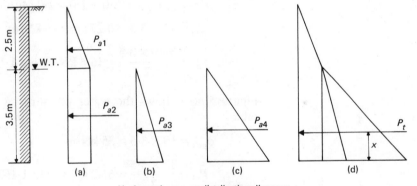

Horizontal pressure distribution diagrams

Figure 5.8

Figure 5.8(*a*) shows the pressure diagram due to the soil which is assumed to be dry above the water table.

Down to the water table

$$\sigma_{a1} = 16 \times 2 \cdot 5 \times 0 \cdot 271 = 10 \cdot 8 \, \text{kN/m}^2$$

and

$$P_{a1} = \frac{10 \cdot 8 \times 2 \cdot 5}{2} = 13 \cdot 5 \, \text{kN acting at } \frac{2 \cdot 5 \times 2}{3} = 1 \cdot 67 \, \text{m from top.}$$

* See footnote p. 2.

Below the water table, this pressure is unchanged and

$$P_{a2} = 10{\cdot}8 \times 3{\cdot}5 = 37{\cdot}8 \text{ kN acting } \left(2{\cdot}5 + \frac{3{\cdot}5}{2}\right) = 4{\cdot}25 \text{ m from top.}$$

Figure 5.8(b) shows the pressure diagram due to the submerged soil below the water table.

Consider 1 m³ of soil.

Void ratio $e = \dfrac{V_v}{V_s} = 0.68$ and $V_v + V_s = 1$

$$\therefore \ V_v + \frac{V_v}{0{\cdot}68} = 1$$

$$\therefore \ V_v = 0{\cdot}40$$

If the voids are saturated

$$W_v = 0.4 \times 1{\cdot}0 = 0{\cdot}4 \, Mg$$
and $$W = 1{\cdot}6 \, Mg$$

$$\therefore \qquad \rho_{\text{sat}} = 2{\cdot}0 \, Mg/m^3$$

$$\therefore \ \rho' = \rho_{\text{sat}} - \rho_w = 2{\cdot}0 - 1{\cdot}0 = 1{\cdot}0 \, Mg/m^3$$

$$\therefore \qquad \sigma_{pa3} = 10 \times 3{\cdot}5 \times 0{\cdot}271 = 9{\cdot}49 \text{ kN/m}^2 \text{ (assuming no change in } \phi)$$

$$\therefore \quad P_{a3} = \frac{9{\cdot}49 \times 3{\cdot}5}{2} = 16{\cdot}6 \text{ kN acting } 1{\cdot}167 \text{ m from bottom.}$$

Figure 5.8(c) shows the water pressure on the wall below the water table.

$$\therefore \ \sigma_{pa4} = 10 \times 3{\cdot}5 = 35 \text{ kN/m}^2$$

$$\therefore \quad P_{a4} = \frac{35 \times 3{\cdot}5}{2} = 61{\cdot}2 \text{ kN acting } 1{\cdot}167 \text{ m from bottom}$$

Figure 5.8(d) shows the combined pressure diagram.

Total thrust on wall

$$P_t = P_{a1} + P_{a2} + P_{a3} + P_{a4}$$
$$= 13{\cdot}5 + 37{\cdot}8 + 16{\cdot}6 + 61{\cdot}2 = \mathbf{129{\cdot}1 \ kN/m \ run}$$

To find where this acts, take moments about the base of the wall of all the forces

$$\therefore \ P_t x = 13{\cdot}5 \times 4{\cdot}33 + 37{\cdot}8 + 1{\cdot}75 + 16{\cdot}6 \times 1{\cdot}167 + 61{\cdot}2 \times 1{\cdot}167$$

$$\therefore \ x = \frac{215{\cdot}4}{129{\cdot}1} = \mathbf{1{\cdot}67 \ m \ from \ base \ of \ wall.}$$

5.4 Active pressure of a cohesive soil on a vertical wall

The soil behind the vertical back of a retaining wall 9 m high consists of soft clay ($c' = 10 \, \text{kN/m}^2$, $\phi' = 20°$, $\rho = 1\cdot60 \, \text{Mg/m}^3$) which is level with the top of the wall. Calculate from first principles the total thrust on the wall assuming that active pressures exist, that tension cracks may develop to the full theoretical depth and that the ground water level may rise to the ground surface.

What is the theoretical depth of the tension crack when the ground surface slopes at inclination i to the horizontal? (ULKC)

Solution The active pressure of a $c - \phi$ soil was shown to be

$$\sigma_{ha} = \gamma z \left[\frac{1 - \sin \phi}{1 + \sin \phi} \right] - 2c \sqrt{\frac{1 - \sin \phi}{1 + \sin \phi}} \qquad \text{[eqn (5.2)]}$$

for the case of a vertical wall and horizontal ground surface. The resulting pressure diagrams are shown on Fig. 5.9.

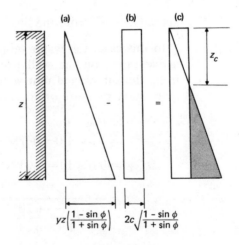

Figure 5.9

Figure 5.9(*a*) is the positive* pressure for the first term (and is the same as for a cohesionless soil).

Figure 5.9(*b*) is the pressure diagram for the second term and is negative. This represents the effect of the cohesion of the soil.

Figure 5.9(*c*) is the resultant pressure distribution diagram.

It can be seen that down to a certain depth z_c the pressure is negative (i.e. tensile) and the soil tends to pull away from the wall. This leads to the formation of *tension cracks* in the soil. The

---

* In soil mechanics compressive forces are considered +ve and tensile forces −ve.

theoretical depth of such a crack z_c can be found by putting $\sigma_{ha} = 0$ in the above equation. Then:

$$\therefore\ \gamma z_c\left(\frac{1 - \sin \phi}{1 + \sin \phi}\right) = 2c\sqrt{\frac{1 - \sin \phi}{1 + \sin \phi}}$$

$$\therefore\ z_c = \frac{2c}{\gamma\sqrt{\dfrac{1 - \sin \phi}{1 + \sin \phi}}}$$

and for a soil where $\phi = 0$, $z_c = \dfrac{2c}{\gamma}$

The total active thrust on the wall will be equal to the area of shaded portion of the pressure diagram Fig. 5.9(c).

$$\therefore\ P_a = \tfrac{1}{2} \times \gamma(z - z_c)\left(\frac{1 - \sin \phi}{1 + \sin \phi}\right) \times (z - z_c)$$

$$= \tfrac{1}{2}\gamma(z - z_c)^2\left(\frac{1 - \sin \phi}{1 + \sin \phi}\right) \tag{5.4}$$

acting at $\dfrac{z - z_c}{3}$ from the base of the wall.

In this case, the water table may rise to the ground surface so the active thrust on the wall from the soil is calculated from eqn (5.4) using the submerged unit weight of the soil. In addition there will be hydrostatic pressure on the full height of the wall.

$$\rho' = \rho_{sat} - \rho_w$$
$$= 1 \cdot 6 - 1 \cdot 0 = 0 \cdot 6 \, \text{Mg/m}^3$$
$$\therefore\ \gamma' = 0 \cdot 6 \times 10 = 6 \cdot 0 \, \text{kN/m}^3 \qquad c' = 10 \, \text{kN/m}^2 \qquad \phi' = 20°$$

$$\therefore\ z_c = \frac{2 \times 10}{6 \cdot 0\sqrt{\dfrac{1 - \sin 20°}{1 + \sin 20°}}} = 4 \cdot 76 \, \text{m}$$

$$\therefore\ P_a = \tfrac{1}{2} \times 6 \cdot 0(9 - 4 \cdot 76)^2\left(\frac{1 - \sin 20°}{1 + \sin 20°}\right) = 26 \, \text{kN/m run}$$

Thrust due to water

$$P_W = \tfrac{1}{2}\gamma_w z^2 = \tfrac{1}{2} \times 10 \times 9^2 = 405 \, \text{kN/m run}$$

$$\therefore\ \text{Total thrust on wall/lineal metre} = \mathbf{431 \, kN.}$$

5.5 Active pressure of two soil layers on a vertical wall

A retaining wall is 7·25 m high. The soil supported consists of 4·5 m of sand ($\gamma = 17 \cdot 5 \, \text{kN/m}^3$, $\phi = 35°$) overlying saturated sandy clay ($\gamma = 19 \cdot 2 \, \text{kN/m}^3$, $\phi = 30°$, $c = 16 \cdot 6 \, \text{kN/m}^2$). The ground-water level is at the upper surface of the sandy clay.

Make a sketch of the distribution of active pressure on the wall, stating the principal values.

Calculate the thrust per lineal metre of wall, neglecting cohesive and frictional forces on the back of the wall. (ICE)

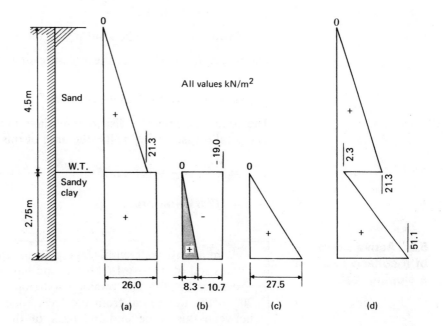

Figure 5.10

Solution Consider 1 m run of wall. Figure 5.10 shows the pressure distribution diagrams on the back of the wall.

Figure 10(*a*) represents the pressure due to the top 4·5 m of sand. When $c = 0$,

$$\sigma_{ha} = \gamma z \frac{1 - \sin \phi}{1 + \sin \phi} = 17\cdot5 \times 4\cdot5 \times 0\cdot271$$

$$= 21\cdot3 \text{ kN/m}^2 \text{ at a depth of } 4\cdot5 \text{ m.}$$

Below this depth the sand layer may be regarded as surcharge on top of the sandy clay and, since ϕ is now 30°,

$$\sigma_{ha} = 17\cdot5 \times 4\cdot5 \times 0\cdot33 = 26\cdot0 \text{ kN/m}^2.$$

Figure 5.10(*b*) represents the pressure due to the sandy clay. Since this is below the water table the submerged density is used in the calculation

$$\sigma_{ha} = \gamma' z \frac{1 - \sin \phi}{1 + \sin \phi} - 2c \sqrt{\frac{1 - \sin \phi}{1 + \sin \phi}}$$

$$= (19 \cdot 2 - 10) \times 2 \cdot 75 \times 0 \cdot 33 - 2 \times 16 \cdot 6 \times \sqrt{0 \cdot 33}$$

$$= 8 \cdot 3 - 19 \cdot 0 = -10 \cdot 7 \text{ kN/m}^2 \text{ at a depth of } 7 \cdot 25 \text{ m below the top of the wall.}$$

At the bottom of the sand $\sigma_{ha} = 26 \cdot 6 - 19 \cdot 0 = 7 \cdot 0 \text{ kN/m}^2$.

Figure 5.10(c) represents the pressure of the water below the water table.

$$\sigma_w = \gamma_w z = 10 \times 2 \cdot 75 = 27 \cdot 5 \text{ kN/m}^2$$

The combined pressure distribution diagram is shown on Fig. 5.10(d). The total thrust on the wall = the area of this diagram

$$= \frac{21 \cdot 3 \times 4 \cdot 5}{2} + 7 \cdot 0 \times 2 \cdot 75 + \frac{36 \cdot 8 \times 2 \cdot 75}{2} = 47 \cdot 9 + 19 \cdot 3 + 50 \cdot 6$$

$$= \mathbf{117 \cdot 8 \text{ kN/m run.}}$$

5.6 Active pressure of a cohesive soil on a sloping wall

> A retaining wall with sloping back, of vertical height 9 m, retains soil of density $1 \cdot 6 \text{ Mg/m}^3$ and having an angle of shearing resistance of 30°. The retained soil slopes upwards uniformly at 20° to the horizontal from the back edge of the wall, the angle between this slope and the back of the wall being 100°. The cohesion of the soil is 10 kN/m^2; the angle of friction between the soil and the wall is 25° but there is no cohesion between them. Find the thrust on the back of the wall per lineal metre using trial values of 25, 30, 40 and 45° for the inclination of the plane of rupture to the vertical.   (UL)

Solution When a retaining wall has a sloping back and the surface of the soil is not horizontal, an analytical solution for the thrust on the wall becomes complex, and a graphical solution based on the wedge theory is adopted.

With the soil in a state of plastic equilibrium, a wedge of soil is about to break away from the main soil mass and is kept from doing so by the forces shown on Fig. 5.11.
The forces which are known are:

W = force due to mass of the soil wedge
C_1 = cohesive force between wedge and main soil mass = $c \times$ length of slip plane
C_2 = cohesive force between wall and soil = $c_w \times$ length of wall in contact with soil.

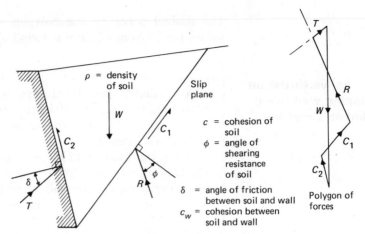

Figure 5.11

The forces for which only the line of action are known are:

   R = reaction between soil mass and wedge acting at ϕ to the normal to slip plane

   T = thrust reaction from wall acting at δ to the normal to wall surface.

By plotting to scale the known forces and adding the lines of action of the other forces a polygon can be drawn and the values of R and T scaled off.

It can be seen that an infinite number of planes could be investigated and the problem is to determine the slip plane which produces the maximum value of T. In order to do this a number of trial planes are chosen and the corresponding polygon of forces drawn, from which the value of T is found.

In the given example a series of planes are chosen as shown on Fig. 5.12, the corresponding values of W, C_1 calculated and R and T found from the force polygon. ($c_w = 0$ in this example.)

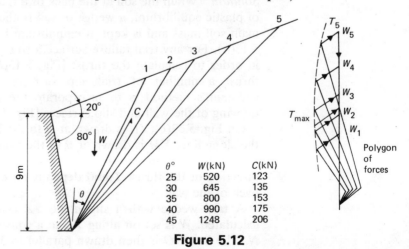

$\theta°$	W(kN)	C(kN)
25	520	123
30	645	135
35	800	153
40	990	175
45	1248	206

Figure 5.12

The dashed curve on the polygon of forces indicates the range of values for T, from which it is found that $T_{max} = 210$ kN/m run.

5.7 Active thrust on a sloping wall using Culmann's method

A retaining wall 4·5 m high with vertical back supports a horizontal fill weighing 19 kN/m³ and having $\phi = 32°$ and $\delta = 20°$ and $c = 0$. Determine the total active thrust on the wall by Culmann's method.

A vertical load of 25 kN is to be carried on a line parallel to the crest of the wall. What is the minimum horizontal distance from the back of the wall at which the load could be placed without increasing the pressure on the wall? (HNC)

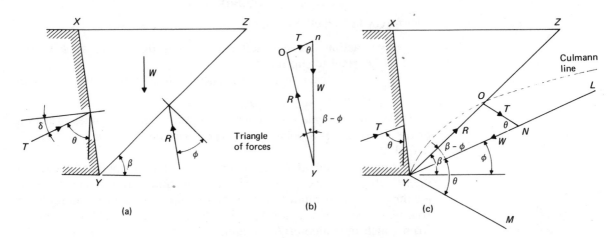

Figure 5.13

Solution When the soil at the back of a retaining wall is just in a state of plastic equilibrium, a wedge of soil is about to break away from the main soil mass and is kept in equilibrium by the forces shown on Fig. 5.13(*a*). For any trial failure surface a triangle of forces may be drawn in order to calculate the thrust [Fig. 5.13(*b*)]. To find the maximum thrust a number of trial slip surfaces have to be investigated. Culmann's method is to incorporate the triangles of forces on the drawing of the wall and slip surfaces [Fig. 5.13(*c*)].

On Fig. 5.13(*c*) *YL* is drawn at ϕ to the horizontal and is known as the slope line. *YM* is drawn at θ to the slope line and is known as the earth pressure line. θ is the angle between the vertical and the direction of the thrust T and depends on δ and the inclination of the back of the wall.

A trial wedge with a slip plane *YZ* is selected and its weight W calculated. W is set off along *YL* to a convenient scale to obtain point N. The line *NO* is then drawn parallel to *YM* to cut the slip plane at

O. The triangle *YNO* thus formed can be seen to be similar to triangle *ynO* of Fig. 5.13(*b*). Thus *NO* represents the thrust *T* from the selected wedge.

This procedure is repeated for other slip planes and the other points corresponding to *O* are joined to form a smooth curve known as the Culmann line. From this line the maximum value of *T* can be found by inspection.

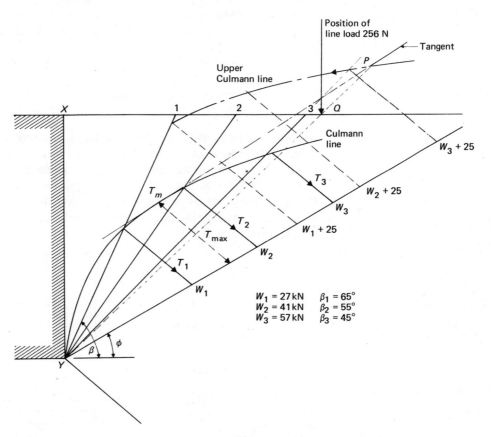

$$W_1 = 27\,\text{kN} \qquad \beta_1 = 65°$$
$$W_2 = 41\,\text{kN} \qquad \beta_2 = 55°$$
$$W_3 = 57\,\text{kN} \qquad \beta_3 = 45°$$

Figure 5.14

The Culmann line for the given wall and soil is given on Fig. 5.14. The maximum thrust on the wall = **20 kN.**

In order to investigate the effect of the line load of 25 kN, it is placed at the top of each slip plane in turn and has the effect of increasing the value of *W* to *W* + 25 kN. The triangle of forces is completed in the same way and a second Culmann line obtained above the initial line.

To find where the line load should be placed in order not to increase the thrust on the wall, a tangent is drawn to the initial Culmann line at the point of maximum thrust T_m to cut the upper line at *P*. A line is then drawn from *Y* to *P* and where this cuts the ground surface *Q* is the closest distance that the line load may be placed to the crest of the wall if the thrust on the wall is not to be increased.

Scaling from Fig. 5.14 the distance $XQ = $ **4·9 m.**

5.8 Stability of a cantilever sheet pile wall in cohesionless soil

An excavation 5·5 m deep in cohesionless soil is supported by a vertical cantilever sheet pile wall. The piling extends to a depth of 3·6 m below the bottom of the excavation. The density of the soil is 1·92 Mg/m³ and $\phi = 33°$. The water table may be assumed to be below the bottom of the piles. Find the thrust on the wall per horizontal foot, neglecting wall friction.

Find also what proportion of the maximum passive resistance is being mobilized on the embedded portion of the piles, stating clearly any simplifying assumptions made. (ICE)

Solution When considering the stability of a cantilever sheet pile wall driven into cohesionless soil, it is usual to assume that at the moment of failure the pile rotates about a point O some distance above the toe (Fig. 5.15). The forces acting on the pile will be the active lateral thrust P_a which tends to cause overturning; the passive resistance P_{p1} acting in front of the pile above O and the passive resistance P_{p2} developed below O.

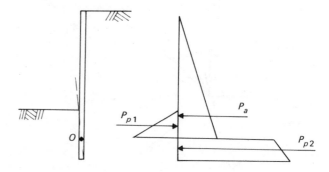

Figure 5.15

To simplify the calculation, the point O is assumed to be at the foot of the pile and the resistance P_{p2} to act as a point load at the same place (Fig. 5.16).

The theoretical depth of penetration required for stability D is then found by taking moments about the toe.

$$P_{p1} \times \frac{D}{3} = P_a \times \frac{(H + D)}{3}$$

The depth D obtained in this way is increased by about 20% to provide a factor of safety.

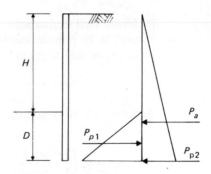

Figure 5.16

In this problem $H = 5\cdot5$ m $\quad \phi = 33°$
$\quad\quad\quad\quad\quad\quad\quad D = 3\cdot6$ m $\quad \rho = 1\,920$ kg/m^3 $\quad \therefore \gamma = 19\cdot2$ kN/m^3
Consider 1 m run of piling:

$$\therefore\ P_a = \tfrac{1}{2}\gamma(H + D)^2\left(\frac{1 - \sin\phi}{1 + \sin\phi}\right)$$

$$= \tfrac{1}{2} \times 19\cdot2 \times 9\cdot1^2\left(\frac{1 - 0\cdot545}{1 + 0\cdot545}\right)$$

$$= \textbf{234 kN} = \text{total thrust on wall/m run.}$$

Taking moments about toe

$$P_{p1} \times \frac{3\cdot6}{3} = P_a \times \frac{9\cdot1}{3}$$

$$\therefore\ P_{p1} = 234 \times \frac{9\cdot1}{3\cdot6} = 591 \text{ kN and represents the passive}$$
$$\text{resistance which must be mobilized}$$
$$\text{for stability of the wall.}$$

The theoretical passive resistance available

$$= \tfrac{1}{2}\gamma D^2\left(\frac{1 + \sin\phi}{1 - \sin\phi}\right)$$

$$= \tfrac{1}{2} \times (19\cdot2) \times 3\cdot6^2 \times \left(\frac{1 + 0\cdot545}{1 - 0\cdot545}\right)$$

$$= 423 \text{ kN which is less than the passive resistance required and}$$
$$\text{the piles must be driven deeper.}$$

5.9 Stability of a cantilever sheet pile wall in a cohesive soil

Derive Bell's equation for active and passive pressure in a $c - \phi$ soil.
A sheet pile wall is driven to a depth z into soft clay which has a bulk density ρ and a cohesive strength c_u ($\phi_u = 0$), and the ground is then excavated from one side of the wall to a depth H.

Using Bell's equations, determine the maximum value of H if the wall is to remain stable. Relate this critical height to that of an unsupported vertical cutting in clay and comment on the result.

(ULKC)

Solution Bell's equations for a $c - \phi$ soil were developed in solution 5.2.

$$\sigma_{ha} = \gamma z \left(\frac{1 - \sin \phi}{1 + \sin \phi}\right) - 2c \sqrt{\frac{1 - \sin \phi}{1 + \sin \phi}} \qquad \text{[eqn (5.2)]}$$

$$\sigma_{hp} = \gamma z \left(\frac{1 + \sin \phi}{1 - \sin \phi}\right) + 2c \sqrt{\frac{1 + \sin \phi}{1 - \sin \phi}} \qquad \text{[eqn (5.3)]}$$

For the case of a cohesive soil ($\phi_u = 0$), these become:

$$\sigma_{ha} = \gamma z - 2c$$

$$\sigma_{hp} = \gamma z + 2c$$

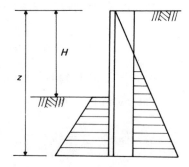

Figure 5.17

Figure 5.17 illustrates the given conditions and shows the active and passive pressure distribution diagrams. The minimum condition for stability in this case is that the total active thrust must be balanced by an equal passive resistance.

$$P_a = \tfrac{1}{2}\gamma z^2 - 2cz$$

$$P_p = \tfrac{1}{2}\gamma(z - H)^2 + 2c(z - H)$$

Equating these values

$$\tfrac{1}{2}\gamma z^2 - 2cz = \tfrac{1}{2}\gamma(z - H)^2 + 2c(z - H)$$

$$\therefore \ \tfrac{1}{2}\gamma z^2 - 2cz = \tfrac{1}{2}\gamma z^2 - \gamma zH + \tfrac{1}{2}\gamma H^2 + 2cz - 2cH$$

$$\therefore \ 2c(H - 2z) = \frac{\gamma H}{2}(H - 2z)$$

$$\therefore \ H = \frac{4c}{\gamma}.$$

This value is the same as that obtained for an unsupported cut in clay (solution 5.1).

Thus, at least initially, no advantage is gained by using sheet piling in this situation.

5.10 Stability of an anchored sheet pile wall using the free earth support method

Explain briefly, with diagrams, the assumptions made in the 'free earth support' method of analysing the stability of anchored sheet piling.

An anchored sheet-pile wall retains soil of height 6·4 m, the piles having a total length of 11 m. The soil has a density of 1·96 Mg/m³, $\phi = 30°$ and the surface is level with the top of the wall. The tie-rods are at 1·2 m below the surface and are spaced 3 m apart horizontally. Neglecting friction on the surface of the piling and assuming 'free earth support', determine:

(*a*) what proportion of the possible passive resistance on the totally embedded length of the piling is mobilized;
(*b*) the pull in the anchor ties.   (UL)

Solution   Cantilever sheet pile walls are not generally very economical if the depth of soil retained exceeds 4·5 m because of the large bending moments induced in the piling and hence the heavy section required.

It is usual, therefore, to introduce horizontal tie bars near the top of the piling. The sheet pile wall is then referred to as an anchored wall.

In the free earth support method of analysing such a wall, it is assumed that the wall is freely supported at the top by the tie rods and at the bottom by the passive resistance of the soil. The general problem is to determine the depth to which the piles should be driven for stability of the wall.

The forces acting are shown on Fig. 5.18.

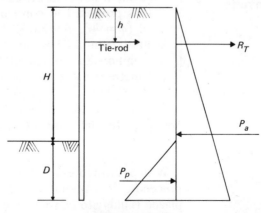

Figure 5.18

In the given case

$$H = 6.4 \text{ m} \qquad D = 4.6 \text{ m} \qquad h = 1.2 \text{ m}$$

$$\gamma = 1.96 \times 10 = 19.6 \text{ kN/m}^2$$

$$P_a = \tfrac{1}{2}\gamma(H + D)^2\left(\frac{1 - \sin \phi}{1 + \sin \phi}\right) = \tfrac{1}{2} \times 19.6 \times 11^2 \times 0.33$$

$$= 391 \text{ kN acting } 7.33 \text{ m from top of wall.}$$

Then taking moments about tie-rod

$$P_p \times 8.27 = P_a \times 6.13$$

$$\therefore \ P_p = 391 \times \frac{6.13}{8.27} = 290 \text{ kN.}$$

This represents the passive resistance required for stability of wall. The theoretical maximum passive resistance of the soil

$$= \tfrac{1}{2}\gamma D^2\left(\frac{1 + \sin \phi}{1 - \sin \phi}\right) = \tfrac{1}{2} \times 19.6 \times 4.6^2 \times 3$$

$$= 622 \text{ kN}$$

$$\therefore \ \text{Proportion mobilized} = \frac{290}{622} = \mathbf{0.47.}$$

For equilibrium Σ horizontal forces $= 0$

$$\therefore \ R_T + P_p = P_a \qquad \therefore \ R_T = 391 - 290 = 101 \text{ kN}$$

If tie rods are 3 m apart

$$\text{Pull} = 101 \times 3 = \mathbf{303 \text{ kN}} \text{ each.}$$

5.11 Pressure distribution on a strutted excavation

Explain why the lateral pressure distribution in a strutted excavation does not increase linearly with depth.

It is proposed to construct a trench 6 m deep in fairly dense sand ($\phi = 40°$, $\rho = 2.0 \text{ Mg/m}^3$) and to timber it with horizontal struts 1.25 m, 2.75 m and 4.5 m below the top. Make reasonable assumptions to estimate the load that each strut will have to carry per metre run of excavation. (CEI)

Solution In the case of a retaining wall, the active pressure is assumed to be mobilized by a rotation of the wall about its base (Fig. 5.3).

With a strutted excavation, the struts are inserted as the excavation proceeds. The amount of excavation before the first row of struts is placed is likely to be small and the original state of stress of the ground almost unchanged.

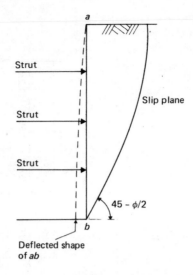

Figure 5.19

As the excavation proceeds the soil tends to yield progressively in a horizontal direction and the resulting deflected shape is shown in Fig. 5.19.

The top strut will prevent the soil from stretching at the start of the sliding wedge and failure will occur as indicated. Thus the lateral pressure distribution will not increase linearly with depth.

The Civil Engineering Code of Practice No. 2—Earth Retaining Structures recommends the pressure distribution shown on Fig. 5.20 for a strutted excavation.

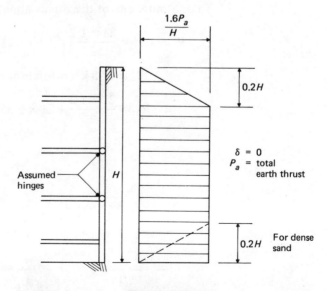

Figure 5.20

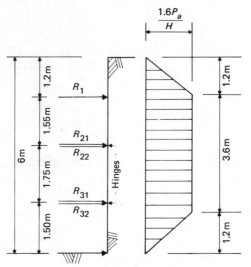

Figure 5.21

Figure 5.21 shows the assumed pressure distribution on the given excavation.

$$P_a = \tfrac{1}{2}\gamma H^2 \frac{1-\sin\phi}{1+\sin\phi} = \tfrac{1}{2} \times (2\cdot0\times10) \times 6^2 \times 0\cdot217$$

$$= 78 \text{ kN/m run of wall}$$

$$\therefore \frac{1\cdot6P_a}{H} = \frac{1\cdot6\times78}{6} = 21 \text{ kN/m}^2.$$

Then the pressure distribution diagrams on each strut, assuming hinges where shown on Fig. 5.21, will be as shown on Fig. 5.22. Taking moments of the forces about R_{21} [Fig. 5.22(a)]

$$R_1 \times 1\cdot55 = \frac{21\times1\cdot2}{2} \times 1\cdot95 + 21 \times \frac{1\cdot55^2}{2}$$

$$\therefore R_1 = \textbf{49\cdot8 kN} = \text{force/m run on top strut}$$

$$\therefore R_{21} = \frac{21\times1\cdot2}{2} + 21 \times 1\cdot55 - 49\cdot8 = 4\cdot65 \text{ kN}$$

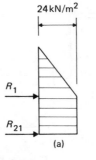

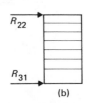

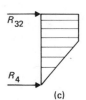

Figure 5.22

From Fig. 5.22(b)

$$R_{22} = R_{31} = \frac{21 \times 1 \cdot 75}{2} = 18 \cdot 4 \text{ kN}$$

$$\therefore \text{ Force/m run on centre strut} = 4 \cdot 65 + 21 = \mathbf{25 \cdot 65 \text{ kN}}.$$

Taking moments about R_4 [Fig. 5.22(c)]

$$R_{32} \times 1 \cdot 55 = 21 \times 0 \cdot 35 \times 1 \cdot 38 + 21 \times \frac{1 \cdot 2}{2} \times 0 \cdot 8$$

$$\therefore R_{32} = 13 \text{ kN}.$$

Total force/m run on lower strut $= 18 \cdot 4 + 13 = \mathbf{31 \cdot 4 \text{ kN}}.$

The reaction R_4 is assumed to be provided by the soil.

5.12 Stability of a vertical trench filled with slurry

A trench of depth H is excavated in a cohesionless soil, angle of shearing resistance ϕ, unit weight γ. The ground-water level is at a height $m . H$ above the level of the base of the trench and the trench is filled to a depth $n . H$ above the level of the base of the trench with a bentonite slurry. If the hydrostatic pressure of the slurry is the only force promoting stability, other than the internal shearing resistance of the soil, show that the minimum slurry unit weight γ_s required for stability is given by:

$$n^2 \frac{\gamma_s}{\gamma_w} = \frac{\gamma}{\gamma_w} K_A + m^2(1 - K_A)$$

where $K_A = \dfrac{1 - \sin \phi}{1 + \sin \phi} = $ Rankine Coefficient of Active Earth Pressure.

Solution Figure 5.23 shows the forces acting at the interface between the slurry and the trench side.

$$P_1 = \tfrac{1}{2}[(1 - m)H]^2 \gamma \frac{1 - \sin \phi}{1 + \sin \phi} = \left(\frac{H^2}{2} - mH^2 + \frac{m^2 H^2}{2} \right) \gamma K_A$$

$$P_2 = (1 - m)H . \gamma . mH . K_A = (mH^2 - m^2 H^2) \gamma K_A$$

$$P_3 = \tfrac{1}{2}(mH)^2 (\gamma - \gamma_w) K_A = \frac{m^2 H^2}{2} \gamma K_A - \frac{m^2 H^2}{2} \gamma_w K_A$$

$$P_4 = \tfrac{1}{2}(mH)^2 \gamma_2 = \frac{m^2 H^2}{2} \gamma_w$$

$P_5 = \tfrac{1}{2}(nH)^2 \gamma_s$, assuming the slurry acts as a liquid of unit weight γ_s.

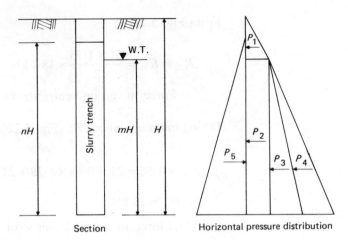

Section   Horizontal pressure distribution

Figure 5.23

For equilibrium $\Sigma H = 0$

$$\therefore\ P_5 = P_1 + P_2 + P_3 + P_4$$

$$\therefore\ \frac{n^2 H^2}{2}\gamma_s = \left(\frac{H^2}{2} - mH^2 + \frac{m^2 H^2}{2} + mH^2 - m^2 H^2 + \frac{m^2 H^2}{2}\right)\gamma K_A$$

$$-\frac{m^2 H^2}{2}\gamma_w K_A + \frac{m^2 H^2}{2}\gamma_w$$

$$\therefore\ n^2 \frac{\gamma_s}{\gamma_w} = \frac{\gamma}{\gamma_w}K_A + m^2(1 - K_A).$$

Problems

1. A retaining wall with vertical back 5·5 m high supports cohesionless fill level with the top. The angle of shearing resistance is 33°. Neglecting the wall friction find the active thrust on the wall/lineal metre:

(*a*) if the soil is well drained (density 1·60 Mg/m³),
(*b*) if the soil is waterlogged (bulk density 1·92 Mg/m³).

Explain briefly how you would find the thrust for case (*a*) taking wall friction into account.   (ICE)

[(*a*) 71 kN; (*b*) 85 kN]

2. A retaining wall with its back vertical retains cohesionless soil to a height of 7·3 m above its base, the surface being horizontal. The angle of shearing resistance is 30°, specific gravity of soil particles 2·65 and void ratio 0·6. The soil is saturated with capillary water but the water table is below foundation level.

Find the direction and magnitude of active thrust per metre run of wall,

(a) neglecting wall friction,
(b) taking angle of wall friction as 25° and assuming that the horizontal component of the active thrust on the wall is 0·83 of the active thrust when the wall friction is neglected.

[(a) 180 kN horizontal; (b) 164 kN at 25° to horizontal]

3. The filling behind a vertical retaining wall 6 m high is cohesionless soil of density 1·76 Mg/m³ and the angle of shearing resistance 35°. The surface is horizontal and level with the top of the wall. In determining the thrust on this wall by the wedge theory, a trial plane of rupture is taken at $27\frac{1}{2}°$ to the vertical. Find the magnitude and direction of the thrust on the wall/lineal foot exerted by the wedge (a) neglecting wall friction; (b) taking the angle of friction on the wall as 25°. (ICE)

[(a) 88 kN horizontal; (b) 75 kN at 25°, to horizontal]

4. A retaining wall of height 9·75 m has the earth face battered at 8 vertical to 1 horizontal. The backfill, which is level with the top of the wall, is clay with density 1·92 Mg/m³, $c = 30$ kN/m² and $\phi = 0$. A trial slip plane is chosen making 35° with the horizontal. Find for this slip plane the thrust per lineal metre taking cohesion between the soil and the wall as $\frac{2}{3}$ the cohesion of the soil. Allow for tension cracks and assume the water table to be below the base of the wall. (ICE)

(360 kN)

5. A retaining wall with sloping back, having a vertical height of 6 m, retains material with a horizontal surface, level with the top of the wall. Warehouses and dockside traffic add a superimposed load which may be taken as 10 kN/m² over the area under consideration. Further details are:

Density of retained material	1·60 Mg/m³
Angle of shearing resistance	30°
Cohesion	10 kN/m²
Angle of friction between material and back of wall	20°
Angle between horizontal ground surface and back of wall	80°

Find the resultant thrust per lineal metre of wall, neglecting the effect of cohesion between the wall and the material. (Suggested trial values of the inclination of the plane of rupture to the vertical: 20°, 25°, 30°, 35°.)

(70 kN)

6. A retaining wall 6 m high supports a fully saturated soil with the following properties: specific gravity of soil particles $G_s = 2.67$, porosity of the soil $n = 0.445$. The surface of the soil is horizontal and level with the top of the wall, and carries a uniform surcharge of 21.5 kN/m^2. Undrained triaxial compression tests on samples of the soil gave the following results:

Lateral pressure (kN/m^2)	7.5	15	18.75
Principal stress difference at failure (kN/m^2)	8.25	10.5	10.1

Assuming that the angle of shearing resistance $\phi = 0$, estimate the cohesion of the soil and determine the magnitude and position of the resultant active pressure on the vertical back of the wall, allowing for the formation of tension cracks at the soil surface. Neglect adhesion and friction along the vertical back of the wall. (UL)

(97 kN/m run, 2 m from base)

7. A sheet-pile wall is to be driven into estuarine clay ($\phi_u = 0$, $c_u = 16.5 \text{ kN/m}^2$, $\rho = 1.76 \text{ Mg/m}^3$) and the ground is to be excavated on one side of the wall to a depth of 6 m leaving 1.75 m of soil to the toe of the piling unexposed. The original ground-water level is 1.75 m below the ground surface.

Assuming that no water content changes take place in the clay and that full active and passive pressures are mobilized, determine the total horizontal thrust using Bell's equations.

How is the thrust likely to alter when the clay swells or consolidates?

Derive the equations used. (ULKC)

(144 kN/m run)

8. A river bank protection scheme consists of a row of sheet piles supported by horizontal tie rods. The vertical height to be retained is 4.5 m. The retained material, including that into which the piles are to be driven, weighs 19 kN/m^3 and has an angle of shearing resistance of 30°. The surface of the retained material is to be horizontal and level with the top of the wall; the tie-rods are to be 1.5 m below this and at 3 m centres measured horizontally. Neglecting cohesion and friction on the surface of the piles, and assuming 'free earth support', find, to the nearest 0.1 m, the minimum length of the piles.

Find also the diameter of the tie-rods, allowing $90 \times 10^3 \text{ kN/m}^2$ tensile stress. (UL)

(2.1 m, 22 mm)

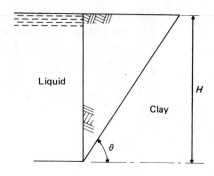

Figure 5.24

9. Explain the use of thixotropic liquids in modern techniques of excavation.

A long trench is being excavated in a homogeneous clay soil having the properties $c_u = 100 \, \text{kN/m}^2$, $\phi_u = 0°$ and $\rho = 1·92 \, \text{Mg/m}^3$. A thixotropic liquid having a density ρ_f is being used to support the cutting. If it is assumed that the liquid deposits a waterproof membrane on the face of the cutting, and that no changes in effective stress take place, a simple theory indicates that the most critical plane to be considered in the ground (Fig. 5.24) is when $\theta = 45°$. Determine the factor of safety against failure of the trench, for which $H = 9 \, \text{m}$, if a fluid with $\rho_f = 1·12 \, \text{Mg/m}^3$ is used. (The factor of safety is defined as $F = c_u/c_r$ where c_r is the cohesion required in limiting equilibrium.) (ICE)

(5.6)

10. A vertical retaining wall supports soil in an active state with a horizontal surface at the top of the wall. Assume that the angle of shearing resistance of the soil is zero and the soil has a cohesive shear strength c whilst the limiting shear adhesion to the wall is nc. Ignoring the possibility of tension cracks show that the angle the critical plane sliding surface makes with the horizontal is

$$\tan^{-1} \frac{1}{\sqrt{1+n}}$$

What is the maximum theoretical depth of a tension crack in this soil? (CEI)

$$\left(\frac{2c}{\gamma}\right)$$

11. A slurry trench is to be excavated in a clay soil of unit weight $20\,kN/m^3$ and average shear strength $40\,kN/m^2$. The slurry is of density $1{\cdot}08\,Mg/m^3$ and fills the trench to ground level. Determine the depth to which the trench can be taken before the factor of safety falls below $1{\cdot}1$. Neglect passive resistance of the slurry. Work from first principles.

($15{\cdot}7\,m$)

6

The stability of soil slopes

When the ground surface is sloping, forces exist which tend to cause the soil to move from high points to low points. The most important of these forces are the force of gravity and the force of seeping water which induce shearing stresses in the soil. Unless the resultant shearing resistance on every plane within the soil mass is greater than the shearing forces, failure will occur in the form of movement of a large mass of soil along a more or less definite surface.

The *stability* of a slope is a measure of its factor of safety against such a failure.

Worked examples

6.1 Factor of safety of a slope against a slide parallel to its surface

Discuss the difference between the 'short-term' and 'long-term' stability of earth structures.

An earth slope is effectively a plane of large extent rising at an angle of β to the horizontal. The long-term failure mode for the slope is likely to be a slide involving a mass of material bounded by a slip plane at a relatively shallow depth z (measured vertically below the surface of the slope) and parallel to the ground surface. The water table is also parallel to the ground surface at a height mz above the slip plane.

Make an effective stress analysis of the slope stability, to show that the factor of safety (with respect to both cohesion and friction) is given by:

$$F = \frac{c' + (\gamma - m\gamma_w)z \cos^2 \beta \tan \phi'}{\gamma z \sin \beta \cos \beta}$$

where c', ϕ' are the shear strength parameters for the soil in terms of effective stress,

γ is the saturated unit weight of the soil, and

γ_w is the unit weight of water.

(ICE)

Solution   All stability analyses are based on the concept that, unless the resultant resistance to shear on every surface is greater than the resultant of all the shearing forces exerted on the surface by the mass above, the slope will fail.

The short-term stability of a slope refers to its stability during and shortly after construction. The shear strength parameters used in the calculations should be the undrained ones c_u and ϕ_u. For long-term stability i.e. some years after construction, the drained values c_d and ϕ_d would be used since the soil will have had time to consolidate. The effect of water seepage and the possible reduction of the shearing resistance of a cohesive soil must also be considered in the long-term stability of a slope.

Consider a prism of soil of unit plan dimensions, within the sliding mass of soil. The forces P on the vertical faces will be in equilibrium (Fig. 6.1).

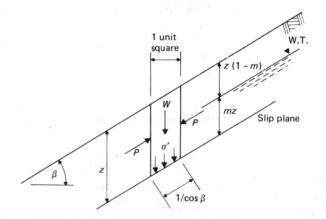

Figure 6.1

The effective vertical stress on the slip plane:

$$\sigma' = \frac{\text{effective weight of prism}}{\text{area of surface}}$$

$$= \frac{\gamma z(1-m) + (\gamma - \gamma_w)mz}{1/\cos \beta} = z(\gamma - m\gamma_w)\cos \beta$$

The component of this stress normal to slip plane

$$\sigma'_n = \sigma' \cos \beta = z(\gamma - m\gamma_w)\cos^2 \beta$$

Substituting in the Coulomb Shear Strength [eqn (4.2)]

$$\tau_f = c' + \sigma'_n \tan \phi' = c' + z(\gamma - m\gamma_w)\cos^2 \beta \tan \phi'$$

$$= \text{resistance of soil to shearing along the slip plane.}$$

The disturbing force which induces shearing stress along the slip plane is the component of the *total* weight of the prism acting along the plane $= \gamma z \sin \beta$

$$\therefore \text{ Disturbing stress} = \frac{\gamma z \sin \beta}{1/\cos \beta} = \gamma z \sin \beta \,.\, \cos \beta$$

The factor of safety $F = \dfrac{\text{Resisting stress}}{\text{Disturbing stress}}$

$$\therefore F = \frac{c' + z(\gamma - m\gamma_w) \cos^2 \beta \tan \phi'}{\gamma z \sin \beta \cos \beta} \tag{6.1}$$

Cohesionless soil slopes

6.2 Stability of a slope in cohesionless soil

Show that a slope of dry cohesionless soil will be stable provided that its angle of inclination to the horizontal β is not more than the angle of shearing resistance ϕ for the soil.

If the slope is subjected to steady seepage and the water table can rise to the ground surface, show how the angle of inclination must be reduced to provide a minimum factor of safety.   (HNC)

Solution   The mode of failure of a dry cohesionless soil will be a slide of material along a shallow slip plane parallel to the ground surface. Thus eqn (6.1) may be applied.

For dry cohesionless soil:

$$c' = 0, \; m\gamma_w = 0, \; \tan \phi' = \tan \phi, \; \gamma = \gamma_d$$

$$\therefore F = \frac{\gamma_d \,.\, z \,.\, \cos^2 \beta \tan \phi}{\gamma_d \,.\, z \,.\, \sin \beta \cos \beta} = \frac{\tan \phi}{\tan \beta}$$

The minimum factor of safety $F = 1$

$$\therefore \text{ For a minimum factor of safety } \tan \beta = \tan \phi.$$

If water is seeping steadily down the slope and the water table is at the ground surface, a flow net indicating these conditions will be as shown in Fig. 6.2.

Consider a prism of soil of depth z and unit plan dimensions measured along the slope.

The drop in head across the prism $\Delta h = \sin \beta$

Width of prism $= 1$

$$\therefore \text{ Hydraulic gradient } i = \frac{\Delta h}{1} = \sin \beta$$

Seepage force/unit volume $j = i\gamma_w$

$$\therefore \text{ Seepage force on prism} = j \times \text{volume} = \sin \beta \,.\, \gamma_w \,.\, z$$

Weight of solid particles in prism $= (\gamma - \gamma_w) \,.\, z$

$$\therefore \text{ Component normal to plane} = (\gamma - \gamma_w)z \cos \beta$$

$$\therefore \text{ Component parallel to plane} = (\gamma - \gamma_w)z \sin \beta$$

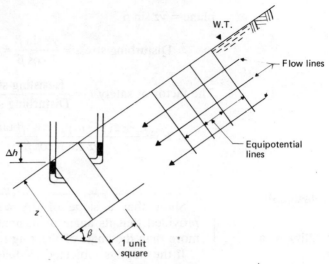

Figure 6.2

$\therefore$ Resisting stress = frictional component of normal stress

$$= (\gamma - \gamma_w)z \cos \beta \tan \phi$$

$$\text{Disturbing stress} = \frac{\text{force parallel to plane} + \text{seepage force}}{\text{area}}$$

$$= (\gamma - \gamma_w)z \sin \beta + \sin \beta \cdot \gamma_w \cdot z$$

$$\therefore F = \frac{(\gamma - \gamma_w)z \cos \beta \tan \phi}{(\gamma - \gamma_w)z \sin \beta + \gamma_w z \sin \beta}$$

$$= \frac{\gamma - \gamma_w}{\gamma} \frac{\tan \phi}{\tan \beta} \quad \text{and for minimum } F = 1$$

$$\therefore \tan \beta = \frac{\gamma'}{\gamma} \tan \phi \quad \text{where } \gamma' = \text{submerged density of soil.}$$

Since γ' is always less than γ, β will be smaller than when the soil is dry, i.e. the slope must be flatter.

Cohesive soil slopes

The stability of a slope in cohesive soil depends on the density, cohesion and angle of shearing resistance of the soil, and the height and inclination of the slope.

It was shown in the previous chapter that it is possible to make vertical cuts in a cohesive soil which will remain stable at least initially.

The most common form of failure of a slope in cohesive soil is a rotational slip along a curved surface which starts beyond the top of the slope and ends at or near the toe. Two types of failure are distinguished:

(1) toe failures which may occur on slopes up to 90° to the horizontal [Fig. 6.3(a)];

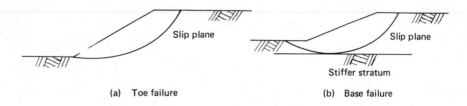

(a)   Toe failure   (b)   Base failure

Figure 6.3

(2) base failures which occur on relatively flat slopes in soft cohesive soil where a harder base exists some distance below the surface [Fig. 6.3(b)].

To analyse the stability of a slope, a potential failure surface is assumed and the forces acting on the surface are calculated.

6.3 Factor of safety against sliding of a slope in cohesive soil

Figure 6.4 represents the section of a clay bank. In an investigation of the bank a trial slip surface *BED* is chosen in the form of a circular arc of radius 11·75 m. The area of the figure *BCDEF* is 87 m² and its centroid is at *G* as shown. The average density of the soil is 1·76 Mg/m³. Above the level of the line *ABE* the cohesion of the soil is 21·5 kN/m², and below this level 33·75 kN/m². Estimate the factor of safety on this assumed surface, taking $\phi = 0°$.

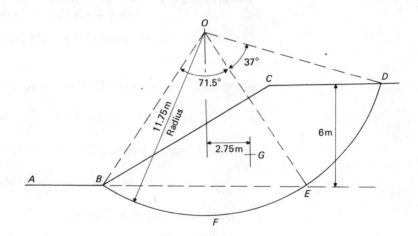

Figure 6.4

Solution $\phi = O$ analysis

Consider a section through the slope of unit thickness. If slipping occurs, it is assumed that it will happen along a circular arc of radius *R*

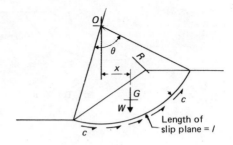

Figure 6.5

and moments are taken about the centre of rotation O of the various forces acting (Fig. 6.5).

The disturbing force will be due to the mass of soil W above the slip surface acting through the centroid of the section G.

$\therefore$ Disturbing moment about $O = Wx$

The resisting force will be the shearing resistance of the soil along the slip surface $= cl$ where l = length of the arc.

$\therefore$ Resisting moment about $O = c \cdot l \cdot R$

and the factor of safety $F = \dfrac{\text{resisting moment}}{\text{disturbing moment}}$

In the given example:

Disturbing force $= W = \gamma \times \text{volume} = (1 \cdot 76 \times 10) \times 87 = 1\,531$ kN
acting vertically through G

Disturbing moment about $O = 1\,531 \times 2 \cdot 75 = 4\,210$ kN m

Resisting forces:

along $BFE = c_1 \times \text{length } BFE = c_1 \times R \times \theta_1$

$$= 33 \cdot 75 \times 11 \cdot 75 \times 71 \cdot 5 \times \frac{\pi}{180} = 495 \text{ kN}$$

along $ED = c_2 \times \text{length } ED$

$$= 21 \cdot 5 \times 11 \cdot 75 \times 37 \times \frac{\pi}{180} = \mathbf{163 \text{ kN}}$$

$\therefore$ Total resisting force $= 658$ kN

$\therefore$ Resisting moment about $O = 658 \times 11 \cdot 75 = 7\,731$ kN m

$\therefore$ Factor of safety $F = \dfrac{7\,731}{4\,210} = \mathbf{1 \cdot 84.}$

In practice a number of trial slip circles would have to be investigated in the same way in order to find the minimum value for F. The one giving the lowest value of F is known as the *critical circle*. Its

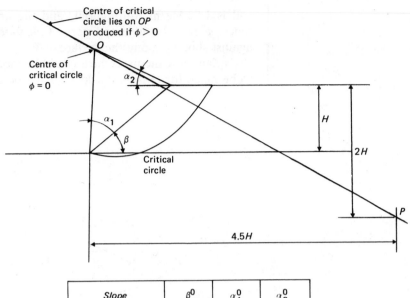

Slope	β^0	α_1^0	α_2^0
1:0.58	60	29	40
1:1	45	28	37
1:1.5	33.8	26	35
1:2	26.6	25	35
1:3	18.4	25	35
1:5	11.3	25	37

Figure 6.6

centre for a homogeneous cohesive soil may be located using Fellenius' construction shown on Fig. 6.6. If $\phi \neq 0$, the centre of the critical surface arc will lie on the line *OP*.

6.4 Factor of safety against sliding of a slope in two layers of soil

Figure 6.7 shows the cross-section of a proposed cutting in cohesive soil. In investigating the stability of the slope a trial slip surface is shown in the form of a circular arc *AB* of radius 18·25 m. The area *ABCD* is found to be 150 m², and its centroid is at *G*, as shown in the sketch. The mean shearing strength of the soil down to a depth of 5·8 m below the top surface is 38·3 kN/m²; below that depth it is 57·5 kN/m². The density of the soil is $1 \cdot 93\,\mathrm{Mg/m^3}$ throughout, and the angle of shearing resistance ϕ may be assumed zero. Calculate the factor of safety against slipping along the surface AB.

Explain how the procedure would be modified if allowance had to be made for tension cracks near the point B. (ICE)

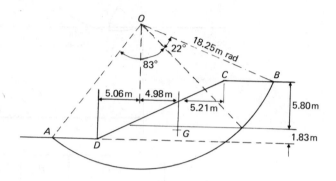

Figure 6.7

Solution This appears to be a base failure rather than a toe failure but the same procedure may be adopted to find the factor of safety since $\phi = 0$.

Disturbing moment about $O = (1 \cdot 93 \times 10) \times 150 \times 4 \cdot 98 = 14\,417\,\mathrm{kN\,m}$

Resisting moment about $O = 57 \cdot 5 \times 18 \cdot 25^2 \times 83\,\dfrac{\pi}{180} + 38 \cdot 3 \times 18 \cdot 25^2$

$$\times \dfrac{22\pi}{180} = 27\,746 + 4\,899 = 32\,645\,\mathrm{kN\,m}$$

$$\therefore \text{ Factor of safety } F = \frac{32\,645}{14\,417} = \mathbf{2 \cdot 37}$$

It was shown in solution 5.4 that it is possible for tension cracks to form in cohesive soils to a depth of $\dfrac{2c}{\gamma}$.

There can be no resistance to shearing along such a crack, so the effect on the analysis of a slip circle is to reduce the length of the arc and hence the resisting force.

In addition the crack may fill with water and the water will exert a horizontal pressure on the soil mass above the slip plane thus adding to the disturbing force (Fig. 6.8).

Thus the presence of a tension crack has the effect of reducing the factor of safety of the slope.

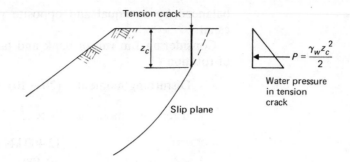

Figure 6.8

6.5 Factor of safety against sliding of a slope in cohesive soil with water level change

The bank of a canal has the profile shown in Fig. 6.9. The material is homogeneous clay of unit weight 20 kN/m^3, cohesion 30 kN/m^2 and angle of shearing resistance zero. For the trial slip circle shown the area $ABCDE$ is 155 m^2 and the centroid is at G. Find for each of the following conditions, the factor of safety for this slip circle:

(a) if the water in the canal is level with the top of the bank,
(b) if the canal is empty.

In both cases allow for a tension crack $2 \cdot 92$ m deep which may be filled with water.   (ICE)

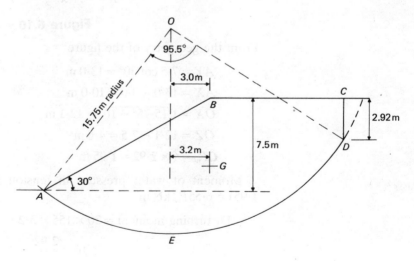

Figure 6.9

Solution (a) When the canal is full of water, the soil may be considered submerged and the submerged density $(\gamma - \gamma_w)$ used in the calculations. The pressure due to water in the tension crack will be balanced by an equal and opposite pressure from the water in the canal.

Considering 1 m run of bank and taking moments about the centre of rotation O.

$$\text{Disturbing moment} = (20 - 10) \times 155 \times 3 \cdot 2 = 4\,960 \text{ kN m}$$

$$\text{and resisting moment} = 30 \times 15 \cdot 75 \times \frac{\pi \times 95 \cdot 5}{180} \times 15 \cdot 75$$

$$= 12\,400 \text{ kN m}$$

$$\therefore \text{ Factor of safety } F = \frac{12\,400}{4\,960} = \mathbf{2 \cdot 5}$$

(*b*) When the canal is empty, the saturated density of the soil is used in the calculations. The pressure of the water in the tension crack adds to the disturbing moment (Fig. 6.10).

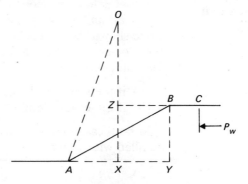

Figure 6.10

From the geometry of the figure:

$$AY = 7 \cdot 5 \cot 30^\circ = 13 \cdot 0 \text{ m}$$

$$AX = 13 \cdot 0 - 3 \cdot 0 = 10 \cdot 0 \text{ m}$$

$$OX = \sqrt{15 \cdot 75^2 - 10^2} = 12 \cdot 1 \text{ m}$$

$$\therefore OZ = 12 \cdot 1 - 7 \cdot 5 = 4 \cdot 6 \text{ m}$$

$$CP_w = \tfrac{2}{3} \times 2 \cdot 92 = 1 \cdot 95 \text{ m}$$

$\therefore$ Moment of water pressure in tension crack about $O = P_w(4 \cdot 6 + 1 \cdot 95) = 6 \cdot 55 P_w$ kN m

Disturbing moment $= 20 \times 155 \times 3 \cdot 2 =$	9 920 kN m
$+\, 10 \times \dfrac{2 \cdot 92^2}{2} \times 6 \cdot 55 =$	280 kN m
	10 200 kN m

Resisting moment $= 12\,400$ kN as before

$$\therefore F = \frac{12\,400}{10\,200} = \mathbf{1 \cdot 21}$$

6.6 Factor of safety against sliding of a slope using stability numbers

Define the term *stability number* as applied to a uniform slope in cohesive soil and derive an expression for it in terms of the cohesion of the soil c, the unit weight γ, the height of the bank H and the factor of safety F.

A cutting of height 9 m in saturated clay soil has a slope of 30° to the horizontal. The properties of the soil are:

ultimate cohesion c   25 kN/m²
angle of shearing resistance ϕ 15°
unit weight γ 20 kN/m³

Find the factor of safety, with respect to cohesion, against a circular slip, assuming friction fully mobilized, if the stability number for these conditions is 0·046.

If friction is not fully mobilized, but the cohesive and frictional forces opposing sliding are assumed to be in the same proportion of their ultimate values, explain how you would find the factor of safety.   (ICE)

Solution   The variable factors which enter into the stability analysis of a uniform slope in cohesive soil are:

(1) apparent cohesion of soil   c kN/m²
(2) angle of shearing resistance of soil ϕ
(3) unit weight of soil γ kN/m³
(4) height of slope   H m
(5) angle of slope β
(6) factor of safety required   F

If four of these variables are combined in the form $\dfrac{c}{F\gamma H}$ the result is a dimensionless number known as a *stability number N*. This has the effect of reducing the variable factors to three and enables charts to be devised relating the stability number to the angle of shearing resistance of the soil and the angle of the slope.

Two slopes are said to be geometrically similar when the ratio of corresponding lengths in the two slopes is constant. From a study of such slopes D. W. Taylor devised a series of curves for the determination of the maximum safe height of a clay slope.

Considering the stability of a unit length of the slope shown on Fig. 6.11.

$$\text{Resisting force} = c \times \text{length of arc } l \propto c \times H$$

$$\text{Disturbing force} = \gamma \times \text{area of sector} \propto \gamma H^2$$

$$\text{and the ratio } \frac{\text{resisting force}}{\text{disturbing force}} = \frac{cH}{\gamma H^2} = \frac{c}{\gamma H}$$

This is the stability number for a factor of safety of 1. To allow for

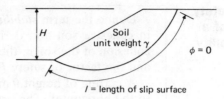

Figure 6.11

different factors of safety the stability number usually appears in the form $N = \dfrac{c}{F\gamma H}$.

For the figures given

$$c = 25\ \text{kN/m}^2,\ N = 0\cdot046,\ H = 9\ \text{m},\ \gamma = 20\ \text{kN/m}^3.$$

$$N = \frac{c}{F\gamma H}$$

$$0\cdot046 = \frac{25}{F \times 20 \times 9}$$

$$F = \textbf{3·0.}$$

The charts have been compiled on the basis of the total stresses in the soil and using Coulomb's shear strength equation

$$\tau_f = c + \sigma \tan \phi$$

The factor of safety is applied only to the cohesive term, assuming that the shearing resistance due to the friction is fully mobilized.

i.e. $\tau = \dfrac{c}{F} + \sigma \tan \phi$

If the factor of safety F is applied to the frictional component also

$$\tau = \frac{c}{F} + \frac{\sigma \tan \phi}{F}$$

and when using the charts a value $\phi_e = \tan^{-1}\!\left(\dfrac{\tan \phi}{F}\right)$ should be used.

In practice the value $\phi_e = \dfrac{\phi}{F}$ is used.

6.7 Factor of safety against sliding of a slope using stability numbers with water level change

A channel with side slopes 1 to 1 has been excavated in cohesive soil to a depth of 5·5 m below ground level. The properties of the soil are

$$\phi = 12°,\ c = 15\ \text{kN/m}^2$$

$$e = 0\cdot9,\ G_s = 2\cdot62$$

Using Taylor's Stability Number $N = \dfrac{c}{F\gamma H}$ find the factor of safety with respect to cohesion against failure of the bank

(a) when the channel is full of water, and
(b) when there is a sudden drawdown of water in the channel.

For case (b) the value of ϕ is reduced to $5 \cdot 6°$.

Data for stability numbers

Slope	Angle of shearing resistance	Stability number
1 to 1	5°	0·136
	10°	0·108
	15°	0·083
		(UL)

Solution

$$\text{Saturated density of soil } \rho_{sat} = \frac{G_s + e}{1 + e} \cdot \rho_w$$

$$= \frac{2 \cdot 62 + 0 \cdot 9}{1 + 0 \cdot 9} \times 1 \cdot 0 = 1 \cdot 85 \text{ Mg/m}^3$$

$$\text{Submerged density of soil } \rho' = \frac{G_s - 1}{1 + e} \cdot \rho_w$$

$$= \frac{2 \cdot 62 - 1}{1 + 0 \cdot 9} \times 1 \cdot 0 = 0 \cdot 85 \text{ Mg/m}^3$$

(a) In this case the submerged unit weight of the soil is used as the water level is at the top of the bank.

When $\phi = 12°$ by interpolating linearly $N = 0 \cdot 098$

$$\gamma = \gamma' = 0 \cdot 85 \times 10 = 8 \cdot 5 \text{ kN/m}^3$$

$$H = 5 \cdot 5 \text{ m}, \qquad c = 15 \text{ kN/m}^2$$

$$N = \frac{c}{F\gamma H}$$

$$\therefore F = \frac{c}{N\gamma H} = \frac{15}{0 \cdot 098 \times 8 \cdot 5 \times 5 \cdot 5} = \textbf{3·27.}$$

(b) With a sudden drawdown, the soil will remain saturated and the saturated density is used in the calculation.

When $\phi = 5 \cdot 6°$ $N = 0 \cdot 133$ by interpolation

$$\gamma = \gamma_{sat} = 1 \cdot 85 \times 10 = 18 \cdot 5 \text{ kN/m}^3$$

$$F = \frac{15}{0 \cdot 133 \times 18 \cdot 5 \times 5 \cdot 5} = \textbf{1·11.}$$

6.8 Side slope angle for a cutting in a cohesive soil

A cutting 12 m deep is to be made in cohesive soil the properties of which are: unit weight 19·2 kN/m³, cohesion 25 kN/m² and angle of shearing resistance $\phi = 15°$. Find a suitable slope for the side of the cutting if the factor of safety against a circular slip is to be 1·5. Assume that the factor of safety applies equally to the cohesive and frictional resistances.

The stability numbers are as follows:

Angle of slope	Stability numbers		
	$\phi = 5°$	$\phi = 10°$	$\phi = 15°$
15°	0·068	0·023	—
30°	0·110	0·075	0·046
45°	0·136	0·108	0·083

Why is a total stress analysis inadequate for estimating long term stability? (ICE)

Solution The stability numbers are compiled on the basis of a factor of safety applied to the cohesive resistance only. To apply the same factor of safety to the frictional resistance, the value of $\phi_e = \frac{\phi}{F}$ is used where $F = $ factor of safety.

$$\therefore \ \phi_e = \frac{15}{1·5} = 10°$$

$$F = \frac{c}{N\gamma H} = \frac{25}{N \times 19·2 \times 12} = \frac{0·1085}{N}$$

Then for 15° slope $N = 0·023$

$$\therefore \ F = \frac{0·1085}{0·023} = 4·7$$

for 30° slope $N = 0·075$

$$\therefore \ F = \frac{0·1085}{0·075} = 1·45$$

for 45° slope $N = 0·108$

$$\therefore \ F = \frac{0·1085}{0·108} = 1·00.$$

These results are plotted on Fig. 6.12 from which when $F = 1·5$ angle of slope = **29°**.

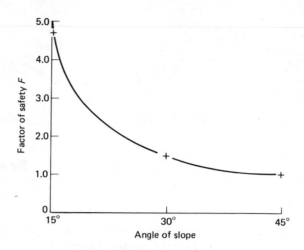

Figure 6.12

6.9 Improvement of factor of safety of a slope against sliding

Explain two ways by which the factor of safety against a circular slip failure of a soil slope can be increased.

The factor of safety of the clay cutting shown on Fig. 6.13 is considered inadequate. In order to increase it, the section is to be altered as shown. Determine the % increase in the factor of safety.

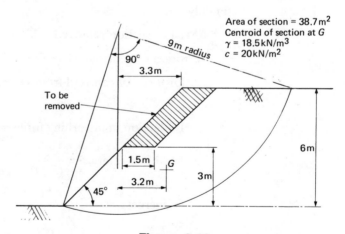

Figure 6.13

Solution The force which causes the slip to occur is the weight of soil in the slope above the slip circle. In order to increase the factor of safety, therefore, the amount of material must be reduced particularly at the top of the slope. Two possible ways of doing this are shown in Fig. 6.14.

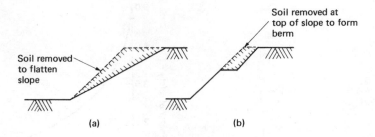

Figure 6.14

In Fig. 6.14(a) the slope of the surface has been flattened and in Fig. 6.14(b) a portion of the top of the slope has been removed. This is a more economical method.

In the given example, initially:

Disturbing force $= 18 \cdot 5 \times 38 \cdot 7 = 716$ kN

$\therefore$ Disturbing moment about $O = 716 \times 3 \cdot 2 = 2\,291$ kN m

Resisting force $= 20 \times \dfrac{2 \times \pi \times 9}{4} = 283$ kN

$\therefore$ Resisting moment $= 283 \times 9 = 2\,545$ kN m

$\therefore$ Factor of safety $F = \dfrac{2\,545}{2\,291} = 1 \cdot 11$

Finally:

Area of section removed $= 1 \cdot 5 \times 3 = 4 \cdot 5$ m^3

Position of centroid of section from $A = \tfrac{1}{2}(3 + 1 \cdot 5) = 2 \cdot 25$ m

$\therefore$ Distance of centroid from vertical through $O = 3 \cdot 3 - (3 - 2 \cdot 25)$
$$= 2 \cdot 55 \text{ m}$$

$\therefore$ Reduction in disturbing force $= 18 \cdot 5 \times 4 \cdot 5 \times 2 \cdot 55 = 212$ kN m

$\therefore$ Factor of safety $F = \dfrac{2\,545}{(2\,291 - 212)} = 1 \cdot 22$

$\therefore$ % increase $= \dfrac{1 \cdot 22 - 1 \cdot 11}{1 \cdot 11} \times 100 = \mathbf{10\%.}$

6.10 Factor of safety of a slope against sliding using method of slices

Using the simplified method of slices derive an expression for the factor of safety (with respect both to cohesion and friction, and in terms of effective stress) for a slope made in material having c' and ϕ' different from zero.

Figure 6.15 shows a slope of total height 3 m in material for which $\phi' = 25°$ and $c' = 0$. The soil is saturated ($\gamma = 19 \cdot 2$ kN/m^3).

Estimate the factor of safety against slip on the trial circle indicated, if the pore pressures are given by $u = 0.2\gamma h$ where h is the mid-height of the slice being considered.　　　(ICE)

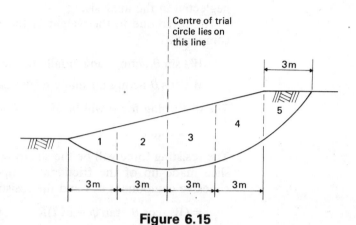

Figure 6.15

Solution　The method of slices is an alternative form of analysis of the stability of a $c - \phi$ soil.

As in other analyses, a circular arc slip plane is assumed. The sector above the failure plane is divided into a convenient number of vertical slices of equal width [Fig. 6.16(b)].

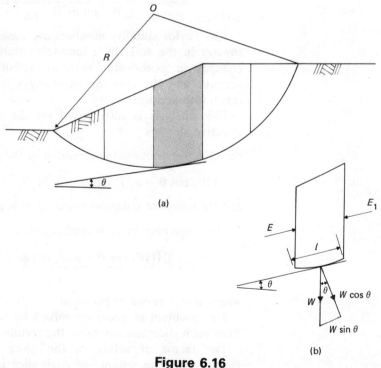

Figure 6.16

Considering the stability of any one strip of unit thickness, the forces acting are shown on Fig. 6.16(b).

Using the simplified method of slices, the forces acting on the sides of the slice E and E_1 are assumed to balance one another and are neglected in the analysis.

The force due to the weight of the slice W can be split up into two components

$W . \sin \theta$ acting tangentially to the base of the slice and

$W . \cos \theta$ acting normally to the base of the slice.

The disturbing force will be $W . \sin \theta$ and its moment about O

$$= W . \sin \theta . R$$

The resisting force will be the shearing resistance along the base of the slice made up of the frictional component $W . \cos \theta . \tan \phi$ and the cohesive component $c . l$ and the resisting moment about O

$$= (W . \cos \theta . \tan \phi + c . l)R$$

The disturbing moment and the resisting moment for the whole section will then be the sum of these moments for each slice.

$$\therefore F = \frac{\Sigma \text{ Resisting moments}}{\Sigma \text{ Disturbing moments}}$$

$$= \frac{\Sigma (W . \cos \theta . \tan \phi + c . l)R}{\Sigma W . \sin \theta . R} = \frac{\Sigma (W . \cos \theta . \tan \phi + c . l)}{\Sigma W . \sin \theta}$$

The Taylor stability numbers are based on the analysis of the *total stresses* in the soil. In a long-term stability analysis of a slope, the changes in pore-water pressure resulting from consolidation and seepage and the consequent changes in effective stresses should be taken into account.

This analysis is also based on the total stresses. Working with effective stresses:

the effective frictional component of the shearing resistance is

$$(W . \cos \theta - u . l) \tan \phi'$$

and the effective cohesive component is $c' . l$

The disturbing force is unchanged.

$$\therefore F = \frac{\Sigma [(W . \cos \theta - u . l) \tan \phi' + c' . l]}{\Sigma W \sin \theta}$$

where $u = $ pore-water pressure.

The problem is normally solved by tabulating the values obtained from each slice and summing the results.

The factor of safety of the given slope may be found semi-graphically. The volume of each slice is approximately equal to the height of the mid-ordinate of the slice multiplied by the width; if this is then multiplied by the unit weight of the soil, it gives W/unit thickness of the slice.

W is assumed to act at the centre of each slice (except for the end ones which are more nearly triangular) and a vertical line representing W is drawn to scale beneath the corresponding slice. A triangle of forces is then drawn, resolving W into its tangential component $(W . \sin \theta)$ and its normal component $(W . \cos \theta)$. Their values are scaled from the drawing and entered on a table together with the value of the pore-water force $(u . l)$. The appropriate columns are summed and the factor of safety found from eqn (6.2).

For the given slope (Fig. 6.17):

width of slice $= 3$ m, unit weight of soil $\gamma = 19\cdot2$ kN/m^3,

pore-water pressure $u = 0\cdot2\gamma h$ where $h =$ height of mid-ordinate.

Slice no.	Height of slice h (m)	Force W $3h\gamma$ (kN)	Tangential component $W . \sin \theta$ (kN)	Normal component $W . \cos \theta$ (kN)	$u . l$ (kN)	$W . \cos \theta - u . l$ (kN)
1	1·07	62	−27	58	13	45
2	2·82	162	−25	161	34	127
3	3·50	202	+30	200	42	158
4	3·35	193	+85	177	40	137
5	1·67	96	+65	71	20	51
Total			+128			518

$$F = \frac{\Sigma\,[(W \cos \theta - u . l) \tan \phi' + c'l]}{\Sigma\,W \sin \theta}$$

$$c' = 0$$

$$F = \frac{518 \tan 25}{128} = \mathbf{1\cdot89}.$$

6.11 Factor of safety of a slope against sliding along composite section

> Estimate the factor of safety of the slope against sliding along the composite section shown on Fig. 6.18 using Bell's equations for the active pressure and passive resistance on a vertical plane.
>
> (HNC)

Solution The presence of a soft stratum below the surface of a slope may lead to a block failure where the slope tends to move horizontally along the soft stratum instead of by rotation along a circular arc. An analysis of the forces causing instability and the resisting forces are shown on Fig. 6.19.

The force tending to cause the slope to move is the active pressure

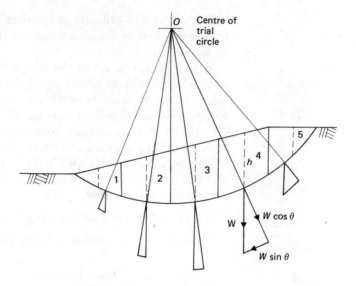

Figure 6.17

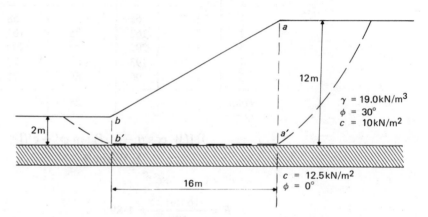

Figure 6.18

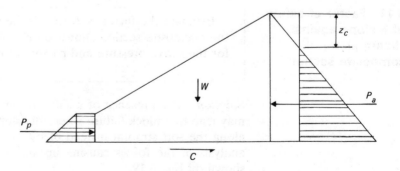

Figure 6.19

P_a of the soil on plane aa'. This is resisted by the passive resistance of the soil P_p on plane bb' and the shearing resistance of the soil along the plane $b'a'$.

The values of P_a and P_b can be found from Bell's formulae (solution 5.2).

$$\sigma_{ha} = \gamma z \left(\frac{1 - \sin \phi}{1 + \sin \phi} \right) - 2c \sqrt{\frac{1 - \sin \phi}{1 + \sin \phi}} \qquad \text{[eqn (5.2)]}$$

$$\sigma_{hp} = \gamma z \left(\frac{1 + \sin \phi}{1 - \sin \phi} \right) + 2c \sqrt{\frac{1 + \sin \phi}{1 - \sin \phi}} \qquad \text{[eqn (5.3)]}$$

For the given case $z_a = 12$ m, $\phi = 30°$, $c = 10$ kN/m^2, $\gamma = 19 \cdot 0$ kN/m^3

$$\therefore \text{ when } \sigma_{ha} = 0, \quad z_c = \frac{2c}{\gamma \sqrt{\dfrac{1 - \sin \phi}{1 + \sin \phi}}} = \frac{2 \times 10}{19 \times \sqrt{\frac{1}{3}}} = 1 \cdot 82 \text{ m}$$

when $z_a = 12$ m $\quad \sigma_{ha} = 19 \times 12 \times \frac{1}{3} - 2 \times 10 \times \sqrt{\frac{1}{3}}$

$$= 76 - 11 \cdot 5 = 64 \cdot 5 \text{ kN/m}^2.$$

$\therefore P_a = $ area of shaded triangle

$$= \tfrac{1}{2} \times 64 \cdot 5 \times (12 - 1 \cdot 82) = 328 \text{ kN/m run of}$$
slope.

$$z_p = 2 \text{ m}$$

$\therefore \sigma_{hp} = 19 \times 2 \times 3 + 2 \times 10 \times \sqrt{3} = 114 + 34 = 148$ kN/m^2

$\therefore P_p = \tfrac{1}{2} \times 114 \times 2 + 34 \times 2 = 182$ kN/m run of slope.

The shearing resistance c is the resistance of the weakest stratum along $b'a'$. For the lower stratum $c = 12 \cdot 5$ kN/m^2, $\phi = 0$

$$\therefore c = 12 \cdot 5 \times 16 = 200 \text{ kN/m run of slope}$$

$$\therefore F = \frac{\text{Resisting force}}{\text{Disturbing force}} = \frac{200 + 182}{328} = \mathbf{1 \cdot 16.}$$

Problems

1. An infinite slope exists at an angle β to the horizontal in a clay soil having a unit weight γ and effective strength parameters c' and ϕ'.

Derive an expression for the factor of safety against failure along a shallow slip plane parallel to the ground surface, and use this to find the maximum stable slope where $c' = 0$, $\phi' = 20°$ and $\gamma = 19$ kN/m^3 assuming that the water table can rise to the ground surface. (CEI)

[see eqn (6.1); 10°]

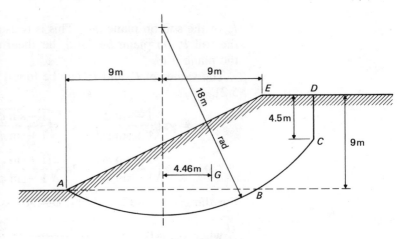

Figure 6.20

2. Figure 6.20 shows the section through a cutting in clay. *ABC* is a trial slip surface and *CD* is an assumed tension crack, 4·5 m deep. The area *ABCDE* is 152 m² and its centroid is at *G*. The density of the soil is 1·92 Mg/m³ and its cohesion is 43 kN/m². Assuming $\phi = 0°$ find the factor of safety against a slip along the surface *ABC*. Allow for the tension crack being filled with water after heavy rain. (UL)

$(F = 1·21)$

3. The conditions of stability of a slope of a uniform soil mass may be expressed by Taylor's Stability Number. Write down the expression for it and indicate by simple reasoning how it is obtained.

A cutting of height 13·75 m is to be excavated in partly saturated clay whose bulk density is 2·0 Mg/m³, cohesion $c = 43$ kN/m² and angle of shearing resistance $\phi = 10°$. Select a suitable slope for the cutting, allowing for a factor of safety 1·7 with respect to both cohesion and friction.

What would be the factor of safety with respect to cohesion against a circular slip of the selected slope, assuming friction fully mobilized?

Data:

Inclination of slope to horizontal	Stability number for	
	$\phi = 5°$	$\phi = 10°$
15°	0·070	0·023
30°	0·110	0·075
45°	0·136	0·108

$(22\frac{1}{2}°; F = 3·2)$ (UL)

4. A wide cutting 6 m deep is to be formed in a non-fissured clay stratum 7·5 m deep which overlies sound chalk. The undrained shear strength of the clay averages 45 kN/m^2 and the density 1·92 Mg/m^3.

Use the stability numbers given in the table for a depth factor of 1·25 to deduce appropriate side slopes if the factor of safety with respect to shear strength is not to be less than 2.

Table of stability numbers for $D = 1·25$

Slope angle (β)	Stability number
10°	0·09
20°	0·13
30°	0·15
45°	0·17
60°	0·19
90°	0·26

Explain how such stability numbers are derived. State whether or not the factor of safety is likely to remain constant with time, giving your reasons, and describe how this can be ascertained.

(ULKC)

5. During the construction of a road embankment on an area of soft normally consolidated soil a slip occurred when the bank had reached a height of 12 m. Subsequent investigation revealed that the slip surface corresponded to that shown in Fig. 6.21 and that at the time of failure pore pressures on the slip surface must have been $u = \gamma h$ where h is the head of soil (of unit weight γ) above the point being considered.

Using the simplified slices method make an estimate of the mean value of ϕ' for the natural soil involved in this slip (adopt the five slices shown).

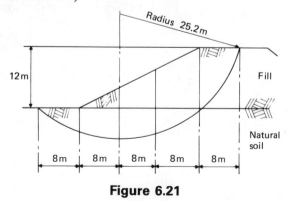

Figure 6.21

What measures might have been taken to avoid the occurrence of this slip?

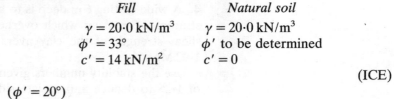

Fill	Natural soil
$\gamma = 20{\cdot}0 \text{ kN/m}^3$	$\gamma = 20{\cdot}0 \text{ kN/m}^3$
$\phi' = 33°$	ϕ' to be determined
$c' = 14 \text{ kN/m}^2$	$c' = 0$

(ICE)

$(\phi' = 20°)$

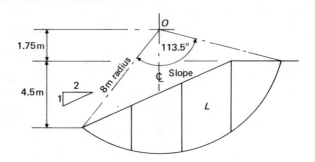

Figure 6.22

6. An embankment has the profile shown in Fig. 6.22. The unit weight of the soil $\gamma = 19{\cdot}2 \text{ kN/m}^3$, its apparent cohesion c' is $6{\cdot}75 \text{ kN/m}^2$ and its angle of shearing resistance ϕ' is $17°$.

(*a*) Assuming tension cracks do not develop, determine the factor of safety with respect to shear strength against failure along the slip circle with centre O. (Use four strips.)

(*b*) There is the possibility of a seepage pressure u developing in the soil mass of the embankment. Assuming that the values of ρ, c' and ϕ' remain unchanged and that $u = 0{\cdot}2\rho gh$, where h is the mid-height of the strip being considered; determine the percentage reduction, due to pore-water pressure, in the factor of safety associated with the same slip circle.   (SCOTEC)

(FOS = 1·5, % reduction = 5%)

7

Bearing capacity of soil and stability of foundations

The ability of a soil to support a load from a structural foundation without failing in shear is known as its *bearing capacity*.

The *stability of a foundation* depends on:

(1) the bearing capacity of the soil beneath the foundation.
(2) the settlement of the soil beneath the foundation.

In order to design a satisfying foundation, there are, therefore, two independent stability conditions to be fulfilled since the shearing resistance of the soil provides the bearing capacity and the consolidation properties determine the settlement.

In order to investigate the bearing capacity of a soil it is necessary first to discover the way in which a shear failure occurs when the bearing capacity is exceeded.

Worked examples

7.1 Bearing pressures on a strip foundation

Give a concise account of the manner in which soil is assumed to fail under shallow foundations. Hence show that in Fig. 7.1 the summation of the components of the passive forces $P_{p\gamma}$, P_{pc}, $P_{pq'}$, give rise to a limiting bearing load Q per unit length of footing where

$$Q = 2P_{p\gamma} \cos(\delta - \phi) + 2P_{pc} \cos(\delta - \phi)$$
$$+ 2P_{pq'} \cos(\delta - \phi) + Bc \tan \delta$$

Hence justify the form of the equation for the bearing pressure on a strip foundation of infinite length:

$$q = \frac{\gamma B}{2} N_\gamma + c \cdot N_c + q' \cdot N_q$$

where c is the cohesion and q' the surcharge.   (ULUC)

Solution A *shallow* footing is one with a breadth $B \geq z$, the depth to the base of the footing. If the footing is infinitely long the problem

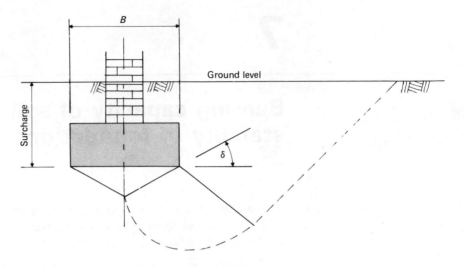

Figure 7.1

may be regarded as two-dimensional, the assumed form of failure is as shown on Fig. 7.2 and no rotation occurs.

A wedge of soil *XYZ* is assumed to move downwards with the footing producing lateral thrusts which, when the soil fails, overcome the passive resistance of the soil along either side of the footing. The form of failure is a slip along the planes *ZHF* and *ZIG*.

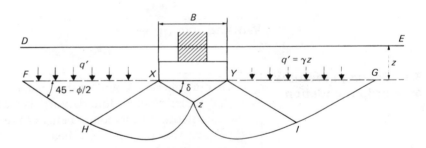

Figure 7.2

In order to analyse this form of failure it is assumed that:

(1) wedges *XHF* and *YIG* are in a state of passive plastic equilibrium,
(2) wedges *XZH* and *YZI* are zones of radial shear,
(3) the soil above the level of the base of the footing does not contribute to the shearing resistance of the soil but acts as a surcharge load.

Consider the freebody diagram showing the forces acting on the wedge of soil under the base of the footing (Fig. 7.3).

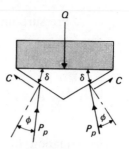

Figure 7.3

Q = load/unit length of footing

C = cohesion along face of wedge = $c\dfrac{B}{2\cos\delta}$

P_p = the total passive resistance force acting at ϕ to the normal.

For the limiting condition when the footing is just in equilibrium and on the point of failing by sinking into the ground:

the sum of all the vertical forces $\sum V = 0$.

$\therefore\ Q = 2C\sin\delta + 2P_p\cos(\delta - \phi)$

The total passive force P_p is made up of three components:

$P_{p\gamma}$ representing the resistance of the wedge XZHF,

P_{pc} representing the cohesive resistance of the soil,

$P_{pq'}$ representing the resistance provided by the surcharge.

Then, substituting for C and P_p, the limiting bearing load/unit length of footing $Q = B\,.\,c\tan\delta + (2P_{p\gamma} + 2P_{pc} + 2P_{pq'})[\cos(\delta - \phi)]$.

For a perfectly smooth base $\delta = 45 + \phi/2$ but in the more likely case of a rough base which prevents the soil at the top of the wedge spreading, $\delta = \phi$.

Then $Q = 2\left(P_{p\gamma} + P_{pc} + P_{pq'} + \dfrac{Bc}{2}\tan\phi\right)$

If the following dimensionless symbols are introduced:

$$N_\gamma = \frac{4P_{p\gamma}}{\gamma B^2}, \qquad N_c = \frac{2P_{pc}}{Bc} + \tan\phi, \qquad N_q = \frac{2P_{pq'}}{q'B}$$

the equation becomes

$$Q = 2\left[\frac{\gamma B^2 . N_\gamma}{4} + \frac{Bc}{2}N_c - \frac{Bc}{2}\tan\phi + \frac{q'BN_q}{2} + \frac{Bc}{2}\tan\phi\right]$$

$$= B\left[\frac{\gamma B}{2}N_\gamma + cN_c + q'N_q\right]$$

$$\therefore \text{ Pressure/unit length } q = \frac{Q}{B}$$

$$\therefore q = \frac{\gamma B}{2} N_\gamma + cN_c + q'N_q$$

where N_γ, N_c and N_q are known as bearing capacity factors.

Shallow foundations The equation for the ultimate bearing capacity of a shallow strip foundation of infinite length developed in solution 7.1 is generally referred to as Terzaghi's equation and may be written in the form:

$$q_f = cN_c + \gamma z N_q + 0 \cdot 5 \gamma B N_\gamma$$

where N_c, N_q and N_γ are dimensionless numbers known as *bearing capacity factors*. Their values depend only on the value of the angle of shearing resistance ϕ for the soil. Graphs of their values have been produced by various workers (e.g. Fig. 7.4). The values of c and ϕ used are normally the undrained values c_u and ϕ_u since these represent the conditions immediately after construction when any failure is most likely to occur; z is the depth below ground level to the base of the footing.

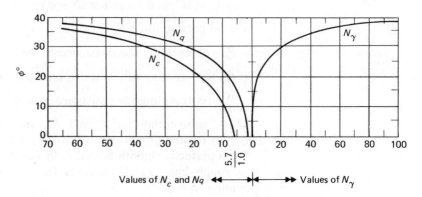

Terzaghi's bearing capacity factors for a shallow footing

Figure 7.4

7.2 Safe bearing capacity of a shallow strip foundation using Terzaghi's equation

Terzaghi's equation for the gross ultimate bearing capacity q_f of a strip footing of breadth B and depth z on soil of cohesion c and unit weight γ is

$$q_f = cN_c + \gamma z N_q + 0 \cdot 5 \gamma B N_\gamma$$

Explain the significance of each of the three terms in this equation.

A strip footing 2·5 m wide is to be constructed at a depth of 3 m below ground level. Find the safe bearing capacity for this footing, using a factor of safety of 3. The cohesion of the soil is 65 kN/m², the density is 1·8 Mg/m³ and the relevant values of the coefficients are $N_c = 10$, $N_q = 4$ and $N_\gamma = 2$. (ICE)

Solution The term cN_c expresses the effect of the cohesion of the soil, $\gamma z N_q$ the effect of the mass of soil above the foundation level and $0.5\gamma B N_\gamma$ the effect of the mass of the wedge of soil on the bearing capacity of the soil. (See solution 7.1.)

From the data given

$$c = 65 \text{ kN/m}^2$$

$$\rho = 1.8 \text{ Mg/m}^3 \qquad \therefore \ \gamma = 1.8 \times 10 = 18 \text{ kN/m}^2$$

$$B = 2.5 \text{ m} \qquad z = 3 \text{ m} \qquad N_c = 10 \qquad N_q = 4 \qquad N_\gamma = 2$$

Substituting in the formula

$$q_f = cN_c + \gamma z N_q + 0.5\gamma B N_\gamma$$

$$\therefore \ q_f = 65 \times 10 + 18 \times 3 \times 4 + 0.5 \times 18 \times 2.5 \times 2 = \textbf{911 kN/m}^2.$$

This is the gross ultimate bearing capacity of the soil.

If a factor of safety of 3 is required the gross safe bearing capacity

$$q_s = \frac{q_f}{3} = \frac{911}{3} = \textbf{304 kN/m}^2.$$

It should be noted that $B < z$ and this is not a shallow footing as previously defined.

7.3 Safe bearing capacity of a shallow footing

Terzaghi's formula for the net ultimate bearing capacity q_{nf} (total pressure less overburden pressure) for a strip footing is

$$q_{nf} = cN_c + \gamma z(N_q - 1) + 0.5\gamma B N_\gamma$$

For a certain soil the cohesion c is 50 kN/m², the unit weight is 19·2 kN/m³ and the coefficients are $N_c = 8$, $N_q = 3$ and $N_\gamma = 2$. Calculate the net ultimate bearing capacity for a strip footing of width $B = 1.25$ m, at a depth $z = 4.5$ m.

Considering shear failure only, calculate the safe total load on a footing 6 m long by 1·25 m wide, using a load factor of 2·5.

What other soil properties should be taken into account in determining the safe load on this foundation? (ICE)

Solution Since the soil at any depth z below the surface is already stressed with the weight of overlying soil ($= \gamma z$ kN/m²), it is usual to base the calculations on the *net* ultimate bearing capacity $q_{nf} = q_f - \gamma z$

which is the pressure additional to the existing pressure which will cause the soil to fail in shear.

$$q_{nf} = 50 \times 8 + 19 \cdot 2 \times 4 \cdot 5(3 - 1) + 0 \cdot 5 \times 19 \cdot 2 \times 1 \cdot 25 \times 2$$
$$= \mathbf{598 \ kN/m^2}$$

The safe bearing pressure with a factor of safety (or load factor) F is then:

$$q_s = \frac{q_{nf}}{F} + \gamma z$$

It should be noted that the strength provided by the overlying soil is always present and does not therefore need to be reduced by a factor of safety.

For a rectangular footing, the formula is modified to take account of the shape, since the problem is no longer just two-dimensional. For a rectangular footing:

$$q_{nf} = cN_c\left(1 + 0 \cdot 3\frac{B}{L}\right) + \gamma z(N_q - 1) + 0 \cdot 5\gamma BN_\gamma\left(1 - 0 \cdot 2\frac{B}{L}\right)$$

where L = length of the foundation.

$$\therefore \ q_{nf} = 50 \times 8\left(1 + 0 \cdot 3\frac{1 \cdot 25}{6}\right) + 19 \cdot 2 \times 4 \cdot 5(3 - 1)$$

$$+ 0 \cdot 5 \times 19 \cdot 2 \times 2\left(1 - 0 \cdot 2\frac{1 \cdot 25}{6}\right)$$

$$= 616 \ kN/m^2, \text{ and with a load factor of } 2 \cdot 5$$

$$q_s = \frac{616}{2 \cdot 5} + 19 \cdot 2 \times 4 \cdot 5 = \mathbf{333 \ kN/m^2}.$$

$\therefore$ Safe load on footing:

$$Q_s = q_s \times \text{area of footing} = 333 \times 6 \times 1 \cdot 25 = \mathbf{2 \ 500 \ kN}.$$

The other property of the soil to be taken into account is its compressibility, since this affects the amount and rate of settlement of the foundation. The allowable bearing pressure q_a must take this into account.

7.4 Comparison of factor of safety of foundation for gross and net pressures

A footing 6 m² square carries a total load, including its own weight, of 10 000 kN. The base of the footing is at a depth of 3 m below the ground surface. The soil strata at the site consist of a layer of stiff fully saturated clay 27·5 m thick overlying dense sand. The average bulk density of the clay is 1·92 Mg/m³ and its average shear strength determined from undrained triaxial tests is 130 kN/m². Determine:

(a) the gross foundation pressure,
(b) the net foundation pressure

and then calculate the factor of safety of the foundation against complete shear failure under undrained conditions. Side cohesion on the foundation may be neglected. (UL)

Solution (a) The gross load on the foundation is the total load from the footing including its own weight = 10 000 kN.

Assuming this is uniformly distributed, the

$$\text{gross foundation pressure} = \frac{\text{load}}{\text{area}} = \frac{10\,000}{6 \times 6} = \textbf{278 kN/m}^2.$$

For a square footing $B = L$ (solution 7.3)

$$\therefore\ q_f = 1 \cdot 3cN_c + \gamma z N_q + 0 \cdot 4\gamma B N_\gamma$$

and from Fig. 7.4 for $\phi = 0$ $N_c = 5 \cdot 7$, $N_q = 1$, $N_\gamma = 0$

$$\therefore\ q_f = 1 \cdot 3 \times 130 \times 5 \cdot 7 + (1 \cdot 92 \times 10) \times 3 \times 1 + 0$$
$$= 1\,021 \text{ kN/m}^2$$

$$\therefore\ \text{Factor of safety} = \frac{1\,021}{278} = \textbf{3} \cdot \textbf{67}.$$

(b) The net foundation pressure = gross foundation pressure − the pressure due to the overburden (removed when the foundation is constructed).

$$\text{Overburden pressure} = \gamma z = (1 \cdot 92 \times 10) \times 3 = 57 \cdot 6 \text{ kN/m}^3$$

$$\therefore\ \text{Net foundation pressure} = 278 - 57 \cdot 6 = 220 \cdot 4 \text{ kN/m}^2$$

$$\therefore\ q_{nf} = 1 \cdot 3 \times 130 \times 5 \cdot 7 + 0 + 0 = 963 \text{ kN/m}^2$$

$$\therefore\ \text{Factor of safety} = \frac{963}{220 \cdot 4} = \textbf{4} \cdot \textbf{37}.$$

7.5 Comparison of factor of safety of foundation with change in water level

A silo, 60 m by 22·5 m in plan, is to be constructed on a slab foundation 3 m below ground level in a uniform clay deposit with an average undrained shearing strength of 75 kN/m². The clay has a saturated density of 1·75 Mg/m³ and the water table, although normally at the ground surface, may rise to 1·5 m above it in times of flooding. If the factor of safety against shear failure is to be not less than 2, determine the maximum uniform vertical load which the silo may carry, assuming the dead weight of the complete structure to be 200×10^3 kN. Allow for soil adhesion on the walls of the silo of 37·5 kN/m².

What is the effect on the factor of safety of the rise of the water level? (ULKC)

Solution With water table at ground level

Weight of overburden removed = vol. of soil excavated $\times \gamma$
$$= 60 \times 22{\cdot}5 \times 3(1{\cdot}75 \times 10)$$
$$= 70{\cdot}9 \times 10^3 \text{ kN}$$

Load carried by side adhesion (with factor of safety of 2)
$$= \frac{\text{area of silo sides} \times \text{adhesion force}}{\text{FOS}} = \frac{2(60 + 22{\cdot}5) \times 3 \times 37{\cdot}5}{2}$$
$$= 9{\cdot}3 \times 10^3 \text{ kN}$$

Buoyancy effect of displaced water = vol. of water displaced $\times \gamma_w$
$$= 60 \times 22{\cdot}5 \times 3 \times 10$$
$$= 40{\cdot}5 \times 10^3 \text{ kN}$$

$\therefore$ Net load on ground at base of foundation

= vertical load Q + dead weight of structure

– overburden removed – load carried by side adhesion

– buoyancy effect of water

$$= Q + 200 \times 10^3 - (70{\cdot}9 + 9{\cdot}3 + 40{\cdot}5)10^3 = Q + 79{\cdot}3 \times 10^3$$

$\therefore$ Net pressure under base assuming uniform pressure distribution
$$= \frac{Q + 79{\cdot}3 \times 10^3}{60 \times 22{\cdot}5}$$

For rectangular base and $\phi = 0°$

$$q_{nf} = cN_c\left(1 + \frac{0{\cdot}3B}{L}\right) = 75 \times 5{\cdot}7\left(1 + \frac{0{\cdot}3 \times 22{\cdot}5}{60}\right)$$
$$= 476 \text{ kN/m}^2$$

$$\therefore \; q_s = \frac{q_{nf}}{F} = \frac{476}{2} = 238 \text{ kN/m}^2$$

$$\therefore \; \frac{Q + 79{\cdot}3 \times 10^3}{60 \times 22{\cdot}5} = 238$$

$$\therefore \; Q = \mathbf{242 \times 10^3 \text{ kN.}}$$

$\therefore$ Q is effectively reduced by $60 \times 22{\cdot}5 \times 1{\cdot}5 \times 10 = 20{\cdot}25 \times 10^3 \text{ kN}$

$$\therefore \; q_{nf} = \frac{(242 - 20{\cdot}25 + 79{\cdot}3) \times 10^3}{60 \times 22{\cdot}5} = 223 \text{ kN/m}^2$$

$$\therefore \; \frac{476}{F} = 223$$

$\therefore$ $F = \mathbf{2{\cdot}13}$ approx., i.e. a slight increase.

Deep foundations

When the depth to the base of a foundation is greater than the breadth of the foundation it may be regarded as a *deep* foundation. The bearing capacity is found using the same approach as for shallow foundations, but the assumption that the weight of soil above the base acts merely as a surcharge becomes increasingly inaccurate and the bearing capacity factors have to be modified.

Myerhof has produced curves for deep strip footings (Fig. 7.5).

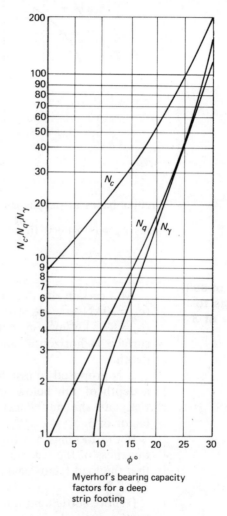

Myerhof's bearing capacity factors for a deep strip footing

Figure 7.5

For a cohesive soil ($\phi_u = 0°$) Skempton has produced curves (Fig. 7.6) showing how the value of N_c varies with the ratio

$$\frac{z}{B} = \frac{\text{depth to foundation level}}{\text{breadth of foundation}}$$

N_c has limiting values of 7·5 for strip footings and 9 for square and

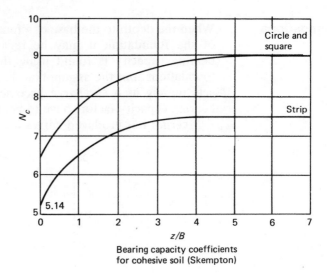

Bearing capacity coefficients
for cohesive soil (Skempton)

Figure 7.6

circular bases at depths exceeding 4 times the width. For rectangular bases $N_c = \left(0\cdot84 + 0\cdot16\dfrac{B}{L}\right) \times N_c$ for a square footing.

7.6 Thickness of soil beneath footings to give load factor of 3

At a certain site the subsoil consists of a thick layer of soft clay $(c_u = 20\,\text{kN/m}^2, \; \phi_u = 0)$ which is overlain by a stiffer clay $(c_u = 87\cdot5\,\text{kN/m}^2, \; \phi_u = 0)$ of variable thickness. The ground surface is horizontal and the water table is at a considerable depth.

It is proposed to install widely spaced footings 1·5 m square at a depth of 1 m below ground level in the upper clay material. Estimate the net working load on the footings, using a load factor of 3.

If the footings are fully loaded at this pressure, estimate the thickness of the stiff clay required below them in order to ensure that the load factor against shear failure in the soft clay is at least 3.

(Ignore settlement effects and assume a load dispersion at 30° to the vertical. Take $N_c = 7\cdot5$ on the lower clay and assume Terzaghi's bearing capacity equations to apply.) (ICE)

Solution Terzaghi's equation for a shallow square footing on a clay soil with $\phi_u = 0$ is:

$$q_{nf} = 1\cdot3cN_c \quad \text{where} \quad N_c = 5\cdot7$$

For upper clay $\therefore q_{nf} = 1\cdot3 \times 87\cdot5 \times 5\cdot7 = 648\,\text{kN/m}^2$

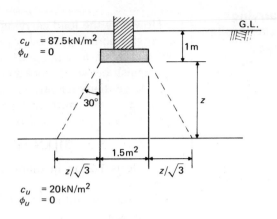

Figure 7.7

∴ using a load factor of 3, the net unit working load

$$= \frac{648}{3} = 216 \text{ kN/m}^2$$

∴ Net working load on footing 1·5 m square

$$= 216 \times 1·5^2 = 486 \text{ kN.}$$

Assuming this is dispersed at 30°, at a depth z below the footing the load will be spread over an area of $\left(1·5 + \frac{2z}{\sqrt{3}}\right)^2 \text{ m}^2$

∴ Pressure on top of lower clay stratum $= \dfrac{486}{\left(1·5 + \dfrac{2z}{\sqrt{3}}\right)^2}$

Allowable pressure $= \dfrac{q_f}{3} = \dfrac{7·5 \times 20}{3} = 50 \text{ kN/m}^2$

Equating these two pressures ∴ $50 = \dfrac{480}{\left(1·5 + \dfrac{2z}{\sqrt{3}}\right)^2}$

∴ $112·5 + 173z + 66·7z^2 = 480$

∴ $z^2 + 2·6z - 5·43 = 0$

∴ $z = \mathbf{1·35}$ **m.**

Pile foundations

One method of transferring a load to a greater depth is by means of a *pile*. A pile is in effect a very slender pier which transfers its load through its lower end to a firm stratum (end bearing pile) or through side friction to the surrounding soil (friction pile) or exerts a combination of both effects.

7.7 Load carrying capacity of single pile

> How may the load carrying capacity of a single driven pile and a single bored pile be assessed? Discuss the differences in behaviour of a single pile and a group of similar piles and distinguish between block failure and local failure of a group.
>
> A large-diameter straight-shafted bored pile is installed in a thick layer of overconsolidated clay whose shear strength at depths z in excess of 3 m below ground level is given by:
>
> $$[75 + 1 \cdot 5(z - 3)] \, kN/m^2$$
>
> The pile is $1 \cdot 25$ m in diameter and has an embedded length of 15 m.
>
> Calculate the ultimate bearing capacity of the pile, stating the shaft adhesion factor used, and making an appropriate allowance for loss of adhesion between pile and clay due to seasonal moisture movements near to the ground surface.
>
> What would you suggest as the working load for this pile? Give your reasons. (ICE)

Solution A driven pile is one which is hammered into the ground and in the process displaces a volume of soil equal to the volume of the pile. A bored pile is one which is formed by the removal of soil by boring.

The load bearing capacity of a single pile may be assessed by considering the measured soil properties adjacent to the pile. A general formula is:

$$Q_f = Q_b + Q_s$$

where Q_f = ultimate load carrying capacity of pile
Q_b = ultimate base resistance due to bearing capacity of ground at toe
Q_s = ultimate shaft resistance due to skin friction over length of pile

For a cohesionless soil:
Q_b may be calculated using Terzaghi's formula for a circular or square footing

$$q_b = p'N_q + 0 \cdot 3\gamma_{sub} \, BN_\gamma$$

where p' is the effective overburden pressure at ground level.

It is generally assumed for deep foundations that the last term is small and may be neglected, that the weight of soil removed is equal to the weight of the pile replacing it and that the water table is at G.L.

$$\therefore \; Q_b = p'N_qA_b \quad \text{where } A_b = \text{area of pile base}$$

Myerhof has suggested that the ultimate skin friction on the side of a pile

$$f_s = K_s p'_m \tan \delta$$

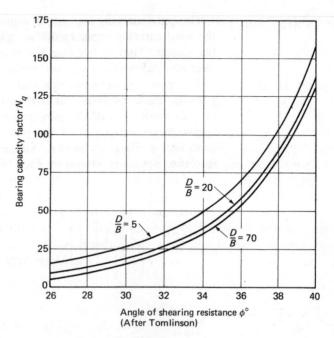

Figure 7.8

where K_s = coefficient of lateral earth pressure acting horizontally on side of pile

p'_m = average overburden pressure on side of pile

δ = angle of friction between soil and pile

$$\therefore\ Q_s = K_s p'_m A_s \tan \delta$$

where A_s = surface area of pile in the ground.

Typical values of K_s are 1·0 for loose sand and 2·0 for dense sand; δ may be taken as 0·75. The values of N_q in Fig. 7.4 are very conservative for piles and the values shown in Fig. 7.8 should be used. For a cohesive soil:

$$Q_b = c_b N_c A_b$$

where c_b = shear strength of soil at base of pile

N_c = bearing capacity factor (usually taken as 9)

$$Q_s = c_s X A_s$$

where c_s = average shear strength of soil adjacent to pile shaft

X = shaft adhesion factor

The most difficult value to assign is often the shaft adhesion factor X. In the absence of particular evidence, values of 0·8 for a driven concrete pile and 0·45 for a bored concrete pile may be assumed.

Both types of pile may be subjected to loading tests, when a maximum allowable settlement is specified under a load of say $1\frac{1}{2}$ times the expected working load.

The load-carrying capacity of a group of piles will not be the same as the load carrying capacity of a single pile times the number of piles in the group. That is because the pressure bulbs for each pile tend to overlap, indicating a greater stress concentration on the surrounding soil. Figure 7.9(a) shows this and also that the influence of a group of piles is more extensive than that of a single pile and this leads to greater settlement of the pile group.

Excessive settlement of the pile group may cause local failure. If, however, a pile group is overloaded, it may fail as a block by breaking into the ground as shown on Fig. 7.9(b).

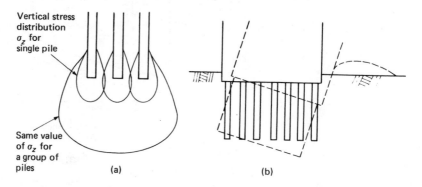

Figure 7.9

For the given pile, the shear strength of the soil adjacent to the shaft is shown on Fig. 7.10.

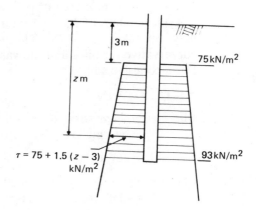

Figure 7.10

Then, neglecting the difference between the weight of the pile and the excavated soil

$$Q_f = Q_b + Q_s$$

where $Q_b = c_b N_c A_b$

$$= [75 + 1 \cdot 5(15 - 3)] \times 9 \times \frac{\pi \times 1 \cdot 25^2}{4} = 1\,027 \text{ kN}$$

$Q_s = c_s X A_s$ and taking the shaft adhesion factor as $0 \cdot 45$

$$= \left(\frac{75 + 93}{2}\right) \times 0 \cdot 45 \times (\pi \times 1 \cdot 25 \times 12)$$

$= 1\,780$ kN neglecting the adhesion along the top 3 m of pile to allow for seasonal moisture movement.

$$\therefore \ Q_f = 1\,027 + 1\,780 = \mathbf{2\,807 \ kN.}$$

The working load on this pile will depend not only on the bearing capacity but also on the permissible settlement. On the basis of the load-carrying capacity alone, using a factor of safety of 3, the working load on the pile is **935 kN**.

7.8 Load carrying capacity of piles using test results

(a) Two independent loading tests on 300 mm diameter short bored piles in clay (for which $\phi_u = 0$) yielded the following results:

Embedded length of pile (m)	Added load at failure (kN)
2·15	100
2·75	110

Assuming the adhesion is effective over the whole of the embedded length, estimate the mean cohesion of the soil and the shaft adhesion factor to be used in extrapolating the test results to larger piles. The densities of the soil and concrete are $1 \cdot 92$ Mg/m³ and $2 \cdot 40$ Mg/m³ respectively.

(b) A bridge pier is to be founded at 4·5 m below bed level in a river estuary. Mean low water level and mean high water level are respectively 3 m and 7·5 m above bed level. The gross load on the pier, which is of circular cross-section, 9 m in diameter, is 30 000 kN; this includes the weight of the foundation.

The clay has a saturated density of $2 \cdot 00$ Mg/m³ and $\phi_u = 0$. Adhesion between the clay and the surface of the pier is estimated at $37 \cdot 5$ kN/m² and is effective over the lower 3 m of pier surface.

Estimate the factors of safety (based on net pressures) against general shear failure at both low and high water. Use a value of $N_c = 7 \cdot 5$. (ICE)

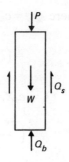

Figure 7.11

Solution (a) 2·15 m pile. At point of failure $\Sigma V = 0$

$\therefore P + (\text{wt. of pile} - \text{wt. of excavated soil}) = Q_b + Q_s$

$$= c_b N_c A_b + c_s X A_s = c N_c \times \frac{\pi}{4} d^2 + c X \pi d l$$

where X = adhesion factor, c = mean cohesion of soil.

$$\frac{z}{d} = \frac{2\,150}{300} = 7 \cdot 16 \quad \therefore N_c = 9 \text{ (from Fig. 7.6)}$$

$$\therefore 100 + (2 \cdot 40 - 1 \cdot 92) \times 10 \left(\frac{\pi}{4} \times 0 \cdot 3^2 \times 2 \cdot 15 \right)$$

$$= c \cdot 9 \cdot \frac{\pi}{4} \times 0 \cdot 3^2 + c X \cdot \pi \times 0 \cdot 3 \times 2 \cdot 15$$

$$\therefore 100 \cdot 729 = 0 \cdot 636 c + 2 \cdot 027 c X \qquad\qquad\qquad \text{(a)}$$

Similarly for 2·75 m pile

$$\therefore 110 + (2 \cdot 40 - 1 \cdot 92) \times 10 \left(\frac{\pi}{4} \times 0 \cdot 3^2 \times 2 \cdot 75 \right)$$

$$= c \cdot 9 \cdot \frac{\pi}{4} \times 0 \cdot 3^2 + c X \cdot \pi \times 0 \cdot 3 \times 2 \cdot 75$$

$$\therefore 110 \cdot 933 = 0 \cdot 636 c + 2 \cdot 592 c X \qquad\qquad\qquad \text{(b)}$$

$$\therefore \text{(b)} - \text{(a)} \qquad 10 \cdot 204 = 0 \cdot 565 c X$$

$$\therefore c X = 18 \cdot 06$$

and substituting in (a)

$$100 \cdot 729 = 0 \cdot 636 c + 2 \cdot 027 \times 18 \cdot 06$$

$$\therefore c = \mathbf{100\ kN/m^2}$$

$$\therefore X = \mathbf{0 \cdot 18}$$

(b) At H.W.L.

Weight of overburden removed

$$= \frac{\pi}{4} \times 9^2 \times 4 \cdot 5 \times 2 \cdot 0 \times 10 = 5\,723 \text{ kN}$$

Load carried by side adhesion $= \pi \times 9 \times 3 \times 37 \cdot 5 = 3\,180 \text{ kN}$

Buoyancy effect of displaced water

$$= \frac{\pi}{4} \times 9^2 \times 12 \times 10 = 7\,630 \text{ kN}$$

∴ Net load on ground at base of pier

$$= 30\,000 - (5\,723 + 3\,180 + 7\,630) = 13\,467 \text{ kN}$$

At L.W.L.

Weight of overburden removed $= 5\,723 \text{ kN}$

Load carried by adhesion $= 3\,180 \text{ kN}$

Buoyancy effect of displaced water $= 7\,630 \times \dfrac{7 \cdot 5}{12} = 4\,768 \text{ kN}$

∴ Net load on ground at base of pier $= 30\,000 - (5\,723 + 3\,180 + 4\,768)$
$$= 16\,329 \text{ kN}$$

$$Q_b = c_b \cdot N_c \cdot A_b$$

If a shaft adhesion factor of 0·45 is assumed

$$c_b = \frac{37 \cdot 5}{0 \cdot 45} = 83 \cdot 3 \text{ kN/m}^2$$

∴ With factor of safety F

At H.W.L.

$$13\,467 \times F = 83 \cdot 3 \times 7 \cdot 5 \times \frac{\pi}{4} \times 9^2$$

$$\therefore F = \mathbf{2 \cdot 95}$$

At L.W.L.

$$16\,329 \times F = 83 \cdot 3 \times 7 \cdot 5 \times \frac{\pi}{4} \times 9^2$$

$$\therefore F = \mathbf{2 \cdot 43}$$

Note: No factor of safety has been allowed for the adhesion. If a factor of safety is applied to this the overall factor of safety is reduced.

What do you understand by the following terms, in relation to piling works: negative skin friction; efficiency and settlement ratio of a pile group; shaft adhesion factor?

A group of concrete piles is square in plan and consists of sixteen piles each 18 m long and 0·45 m in diameter. They are installed at 1·10 m centres in a deep deposit of estuarine clay having a mean unconfined compressive strength over the pile length of $c_u = 30$ kN/m^2. At and below the level of the pile bases the mean strength is $c_u = 40$ kN/m^2. The mean unit weights of the soil and concrete are respectively 19·2 kN/m^2 and 24·0 kN/m^2.

Stating clearly your assumptions concerning bearing capacity and shaft adhesion factors estimate the superimposed load required to cause failure of a single pile if installed (*a*) by boring (*b*) by driving.

What total working load would you recommend for the pile group immediately after its installation by driving?   (ICE)

Solution Negative skin friction: in general, friction between the pile shaft and the surrounding soil tends to increase the load carrying capacity of the pile. In special circumstances, such as piles driven through soil which tends to consolidate under its own weight or compressible fills, a drag-down force will act on the pile and reduce its bearing capacity. Such a force is known as negative skin friction.

$$\text{Efficiency ratio} = \frac{\text{Average load per pile in group at failure}}{\text{Failure load of single comparable pile}}$$

$$\text{Settlement ratio} = \frac{\text{Settlement of pile group}}{\text{Settlement of single pile}}$$

$$\text{Shaft adhesion factor} = \frac{\text{Skin friction on shaft of pile}}{\text{Shear strength of soil adjoining shaft}}$$

(*a*) single bored pile:

$$Q_f = c_b N_c A_b + c_s X A_s \qquad \text{(see solution 7.8)}$$

and using a shaft adhesion factor of 0·45

$$\therefore Q_f = 40 \times 9 \times \left(\frac{\pi}{4} \times 0\cdot45^2\right) + 30 \times 0\cdot45 \times (\pi \times 0\cdot45 \times 18)$$

$$= 57 + 343 = 400 \text{ kN}$$

If the difference in weight between the pile and the displaced soil is taken into account this figure is reduced by

$$(24\cdot0 - 19\cdot2) \times \frac{\pi}{4} \times 0\cdot45^2 \times 18 = 14 \text{ kN}, \qquad \text{i.e. } 3\cdot4\%.$$

(b) single driven pile:

A shaft adhesion factor of 0·8 may be used,

$$Q_f = 57 + 343 \times \frac{0·8}{0·45} = 667 \text{ kN}$$

It can be seen that the difference between the weight of the pile and the weight of the displaced soil has a negligible effect but the choice of the shaft adhesion factor is critical.

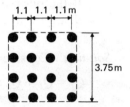

Figure 7.12

The group of 16 piles are shown on Fig. 7.12. In order to calculate the ultimate load on a group of piles it is usual to treat it as a footing with dimensions equal to the overall dimensions of the pile group. On this basis:

$$\begin{aligned} Q_{fg} &= c_b N_c A_b + c_s X A_s \\ &= 40 \times 9 \times 3·75^2 + 30 \times 0·8 \times (4 \times 3·75 \times 18) \\ &= 5\,060 + 6\,480 = 11\,540 \text{ kN} \end{aligned}$$

Terzaghi and Peck state that a group of piles will be safe against a collapse failure if the number of piles times the safe design load per pile

$$< \frac{Q_{fg}}{3}$$

∴ Maximum load on group on this basis $= \dfrac{11\,540}{3} = 3\,847 \text{ kN}$

∴ Factor of safety on an individual pile $= \dfrac{667 \times 16}{3\,847} = 2·77$

7.10 Number of piles required to carry a column load

It is proposed to carry the load from a column on 350 mm diameter bored piles. The total load to be carried is 2 000 kN and the soil consists of a deep deposit of clay having the following properties:

Depth (m)	Undrained shear strength (kN/m²)
0	100
3	110
6	125
9	125
12	135

If the piles are about 9 m long, estimate the number required and suggest how they should be arranged.　　　　(ULKC)

Solution　Consider the ultimate load carried on pile.

$Q_f = Q_b + Q_s$ and taking average c_u on shaft of pile as $= 115$ kN/m²
and an adhesion factor of 0·45 for a bored pile.

$$Q_f = 125 \times 9 \times \left(\frac{\pi}{4} \times 0·35^2\right) + 115 \times 0·45 \times (\pi \times 0·35 \times 9)$$

$$= 108 + 512 = 620 \text{ kN}$$

Using a factor of safety of 3:　$Q_a = \dfrac{620}{3} = 207$ kN

$\therefore$ No. of piles required to carry 2 000 kN $= \dfrac{2\,000}{207} = 9·6$

say 9 piles which reduces the factor of safety slightly.

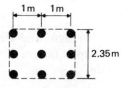

Figure 7.13

Bored piles should be spaced at about 3 times their diameter and a suggested arrangement is shown on Fig. 7.13.

The action of the group must be checked for stability.

Consider a square pier 2·35 m square

$$\therefore Q_{fg} = 125 \times 9 \times 2·35^2 + 115 \times 0·45 \times (4 \times 2·35 \times 9)$$

$$= 6\,212 + 4\,378 = 10\,590 \text{ kN}$$

$$\therefore Q_{fg} = \frac{10\,590}{3} = 3\,530 \text{ kN}$$

Since this is more than the load of 2 000 kN, the arrangement should be satisfactory from a load carrying point of view.

If the piles had been spaced at $1 \cdot 5 \times$ dia the block would measure $(0 \cdot 35 \times 1 \cdot 5 \times 2) + 0 \cdot 35 = 1 \cdot 4$ m each side

$$\therefore Q_{fg} = 125 \times 9 \times 1 \cdot 4^2 + 115 \times 0 \cdot 45 \times (4 \times 1 \cdot 4 \times 9)$$
$$= 4 \, 813 \text{ kN}$$

$$\therefore Q_{fa} = \frac{4 \, 813}{3} = 1 \, 604 \text{ kN which is inadequate.}$$

$\therefore$ spacing should be at least $2 \times$ diameter of piles.

7.11 Length of pile with under-reamed base to carry column load

A bored pile with an under-reamed base in a deep deposit of clay is required to carry a load of 1 000 kN. The diameter of the shaft is to be 0·75 m and the undrained shear strength of the soil $c_u = 95$ kN/m². Determine the length of the pile and the diameter of the base if the side slope of the under-reamed section is 1 horizontal to 3 vertical. Estimate the settlement that will occur.

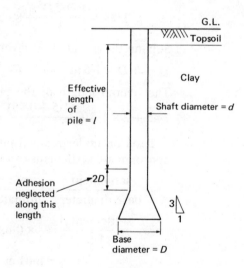

Figure 7.14

Solution   The pile is shown on Fig. 7.14, and as with other piles $Q_f = Q_b + Q_s$.

$Q_b = c_b \cdot N_c \cdot A_b$ where $c_b = c_u = 95$ kN/m² and N_c is taken as $= 9$

$Q_s = c_s \cdot X \cdot A_s$ where $c_s = c_u = 95$ kN/m².

In the case of a large bored pile with an under-reamed base, the time required for construction allows seepage to occur along the shaft with consequent weakening of the clay. An adhesion factor $X = 0 \cdot 3$ is generally used. It is also usual to neglect adhesion for a length of two base diameters above the tapered portion.

$$\therefore Q_b = 95 \times 9 \times \frac{\pi}{4} \times D^2 = 671 \cdot 5D^2$$

$$\therefore Q_s = 95 \times 0 \cdot 3 \times \pi \times d \times l = 89 \cdot 5dl$$

$$\therefore Q_f = 671 \cdot 5D^2 + 89 \cdot 5dl$$

It has been found for this type of pile that the full shaft resistance is mobilized after a settlement of about 15 mm but that the base resistance is not mobilized until the settlement reaches about 120 mm. The safe load Q_s is generally taken as the lesser of:

$$Q_s = \frac{Q_f}{2} \quad \text{or} \quad Q_s = \frac{Q_b}{3} + Q_s$$

The difference between the weight of the pile and the weight of excavated soil is relatively small and can be neglected, and thus

$$Q_s = 1\,000 \text{ kN}$$

$$\therefore 1\,000 = \frac{671 \cdot 5D^2 + 89 \cdot 5dl}{2}$$

$$\therefore 1\,000 = \frac{671 \cdot 5D^2}{3} + 89 \cdot 5dl$$

Solving these equations given that $d = 0 \cdot 75$ m

$$\therefore D = 1 \cdot 5 \text{ m} \quad \text{and} \quad l = 7 \cdot 45 \text{ m}$$

The overall length of the pile will therefore be = depth of topsoil +

$$7 \cdot 45 + 2 \times 1 \cdot 5 + \frac{(1 \cdot 5 - 0 \cdot 75)}{2} \times 3 = \textbf{depth of topsoil} + \textbf{11} \cdot \textbf{57 m.}$$

Tests on under-reamed piles have shown that in uniform clays, the approximate settlement can be calculated using the empirical formula:

$$\frac{\text{settlement}}{\text{base diameter}} = \frac{\text{safe load on base}}{\text{ultimate load on base}} \times 0 \cdot 02$$

$$\therefore \frac{\text{settlement } \rho}{1 \cdot 5} = \frac{1}{3} \times 0 \cdot 02$$

$$\rho = \textbf{0} \cdot \textbf{01 m}$$

7.12 Factor of safety against base failure of a strutted excavation

Explain the significance of bearing capacity factors and illustrate their uses. A strutted excavation 6 m $\times$ 7·5 m in plan is to be taken to a depth of 6 m in a sensitive clay which has an undisturbed shear strength (c_u) of 15 kN/m^2 and a density (ρ) of 1·76 Mg/m^3. The ground surface carries an overall surcharge of 10 kN/m^2.

What is the factor of safety against a base failure?

(ULKC and CEI)

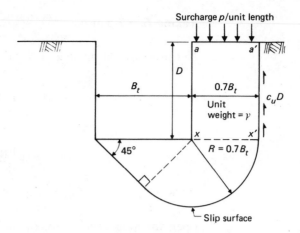

Figure 7.15

Solution The significance and uses of bearing capacity factors have been discussed in this chapter.

Figure 7.15 shows a strutted excavation and the generally assumed mechanism of a base failure in a cohesive soil. The slope of the straight portion of the slip surface is at 45° to the horizontal and the curved portion is a circular arc with its centre at the base of the side sheeting. From the geometry of the figure, the radius of this arc $= B_t$. $\sin 45° = 0.7B_t$.

Surface $x-x'$ may be regarded as the base of a footing carrying the weight of the soil above it plus any surcharge on the surface $a-a'$. The downward movement is resisted by the shearing resistance along the surface $a'-x'$. Considering a unit length of excavation.

$$\therefore \text{ Net load on plane } x-x' = 0.7B_t D\gamma - c_u D + 0.7B_t p$$

The net bearing capacity is calculated in the same way as for a footing, using appropriate bearing capacity factors.

$$\therefore q_{nf} = c_u N_c \text{ for a strip footing}$$

$$= c_u N_c\left(1 + 0.3\frac{B}{L}\right) \text{ for a rectangular footing.}$$

For the given conditions: Taking $B_t = 6\,\text{m}$, $D = 6\,\text{m}$

$$\text{Net load on plane } x-x' = (0.7 \times 6) \times 6 \times (1.76 \times 10) - 15 \times 6$$
$$+ (0.7 \times 6) \times 10 = 396\,\text{kN}$$

$$\therefore \text{ net pressure } = \frac{396}{0.7 \times 6} = 94.3\,\text{kN/m}^2$$

$$q_{nf} = 15 \times 5.7\left(1 + \frac{0.3 \times 0.7 \times 6}{7.5}\right) = 100\,\text{kN/m}^2$$

$$\therefore \text{ Factor of safety } F = \frac{100}{94.3} = 1.06$$

Taking $B_t = 7.5$ m

Net pressure on plane x–$x' = 6(1.76 \times 10) - \dfrac{15 \times 6}{0.7 \times 7.5} + 10 = 98$ kN/m^2

$$q_{nf} = 15 \times 5.7\left(1 + \frac{0.3 \times 0.7 \times 7.5}{6}\right)$$

$$= 108 \text{ kN}$$

$$\therefore F = \frac{108}{98} = 1.10$$

Thus the excavation is just safe from a base failure and this may be satisfactory if the excavation is temporary and not likely to be left open for long. The factor of safety could be increased by driving the side sheeting lower than the depth of excavation.

Problems

1. Explain what is meant by the ultimate bearing capacity of a soil. How is the ultimate bearing capacity of a cohesive soil related to its shearing strength?

A reinforced concrete column has a square footing founded at a depth of 3 m below ground level on clay of average density 1.76 Mg/m^3 and shearing strength 36 kN/m^2. The total load applied to the soil is 800 kN. Calculate the dimensions of the footing assuming a suitable factor of safety. (ICE)

(3 m × 3 m with FOS = 3)

2. A strip footing 3.5 m wide is to be placed 3 m below ground surface on sandy clay having a bulk density ρ of 2.05 Mg/m^3. Undrained shear box tests give shear strengths of 35, 47 and 59 kN/m^2 for normal stresses of 70, 140 and 210 kN/m^2 respectively.

Find the apparent cohesion c and angle of shearing resistance ϕ.

Calculate the ultimate load/m run of foundation immediately after construction using Terzaghi's formula. The bearing capacity coefficients may be taken as $N_c = 10$, $N_q = 4$ and $N_\gamma = 2$. (ICE)

($c = 25$ kN/m^2, $\phi = 9°$; 1988 kN)

3. A square footing is to be constructed at a depth of 3.6 m below ground surface on sandy clay soil for which the cohesion c is 57.5 kN/m^2 and the density ρ is 1.76 Mg/m^3. The total load applied to the soil is 4300 kN, uniformly distributed over the area of contact. Find the size of the footing, using Terzaghi's formula for the net ultimate bearing capacity:

$$q_u = 1 \cdot 3cN_c + \gamma z(N_q - 1) + 0 \cdot 4\gamma BN_\gamma.$$

Use a load factor of 3 and take the relevant values of the coefficients as $N_c = 10$, $N_q = 4$, $N_\gamma = 2$. (UL)

($3 \cdot 61$ m)

4. A square footing is required to carry a total load of 2 500 kN. The base of the footing is to be $3 \cdot 6$ m below ground level. The soil is saturated clay having the following properties:

Apparent cohesion c	60 kN/m^2
Angle of shearing resistance	0°
Density	$1 \cdot 92$ Mg/m^3

The ultimate bearing capacity of the soil for a surface load may be taken as $6c$. Assuming uniform pressure over the area of the base and using a factor of safety of 3 against shear failure, find a suitable size for the footing.

The clay extends to a depth of $9 \cdot 75$ m below ground level and below this there is an incompressible stratum. Assuming that the mean vertical pressure in the clay under the footing is $0 \cdot 75$ of the contact pressure between the base of the footing and the soil, and taking the coefficient of volume decrease m_v as 112×10^{-6} m^2/kN estimate the final settlement of the footing. (UL)

($1 \cdot 4$ m square, $0 \cdot 66$ m)

5. A square footing is required to support a vertical load, uniformly distributed, of 1 500 kN, on clay soil for which the cohesion c is $57 \cdot 5$ kN/m^2, the unit weight $= 19 \cdot 2$ kN/m^2 and $\phi = 0$. The base of the footing is to be at $4 \cdot 5$ m below ground level. Find a suitable size for the footing, using a load factor of $2 \cdot 5$ and taking the net ultimate bearing capacity as $7 \cdot 5c$.

The clay extends to a depth of $3 \cdot 6$ m under the base of the footing, and below this level there is a stratum of relatively incompressible material. The average pressure increment over this $3 \cdot 6$ m depth due to the applied load may be taken as $0 \cdot 56$ of the contact pressure under the footing. Consolidation tests on the clay under similar conditions of pressure distribution and drainage show that a pressure increment of 10 kN/m^2 causes a settlement of 4% of the thickness of the sample. Find the probable settlement of the footing. (UL)

($2 \cdot 4$ m^2, $1 \cdot 4$ m)

6. Examine critically the main methods employed for the assessment of the safe bearing capacity of shallow footings.

Over much of a certain site levelling has been achieved by the introduction of a large depth of fill material having average properties:

$$\phi_u = 20°, \ c_u = 28 \text{ kN/m}^2, \text{ bulk density } 2·08 \text{ Mg/m}^3.$$

Over the rest of the site the foundation material is a natural clay having the average properties: $\phi_u = 0°$, $c_u = 100 \text{ kN/m}^2$. The water table is normally at a considerable depth.

It is proposed to install strip footings at a depth of 1·5 m below ground level. Determine the width of the footing required in each type of soil if the net loading intensity (after allowing for the concrete in the footing and backfilling to ground level) is to be 278 kN/m run. A factor of safety of 3 is required against the possibility of general shear failure.

Describe what redesign would be necessary if the site became waterlogged, with the water table near to the surface. For a soil with $\phi = 20°$ the Terzaghi bearing capacity factors are: $N_c = 17$, $N_q = 7$, $N_\gamma = 5$. (ICE)

(On fill $B = 1·26$ m, on clay $B = 1·46$ m)

7. A mass concrete pier of circular section is to carry a load, including its own weight, of 2 500 kN. The base of the pier is to be located 4·5 m below the surface of a deep layer of saturated clay. The average bulk density of the clay is 1·84 Mg/m^3 and its average undrained shear strength $c_u = 50 \text{ kN/m}^2$. Choosing a suitable value for the bearing capacity factor N_c determine the diameter of the pier to give a factor of safety of 3 against a complete foundation failure. Assume that a cohesion $0·3c_u$ can be mobilized on the cylindrical surface of the pier below ground level. (LU)

(1·52 m taking $c_u = 7·5$)

8. A group of 9 bearing piles of 300 mm diameter, is driven in a square grid with the lines of the piles at 900 mm centres. The piles are cast *in-situ* and are 6·5 m long. The soil is a deep deposit of clay having the following properties:

$$c_u = 72 \text{ kN/m}^2$$
$$\phi_u = 0°$$
$$\rho = 2·00 \text{ Mg/m}^3$$

Determine the bearing capacity of the pile group, if the minimum factor of safety is 3.

(2 000 kN)

9. A series of undrained triaxial tests on clay taken from the site of a proposed building in central London gave the following results:

	O.D. level	Shearing resistance (kN/m²)	Bulk density (Mg/m³)
Ground surface	+6 m	—	2·08
	0	165	1·92
	−6	210	1·95
	−12	250	1·98
	−18	295	2·02
	−24	330	2·05

The top 4·5 m is Thames Ballast. Estimate the load which a bored pile of 1·40 m diameter and 30 m in depth below ground surface can carry in this clay if the bottom is belled out at an angle of 60° to the horizontal to a diameter of 3·5 m.

(13 930 kN)

8

Compaction and stabilization of soil

When a soil is capable of continuous resistance to lateral displacement under load, it is said to be stable. This is particularly important when soil is used as foundation material for a road or as construction material for an embankment.

The properties of a soil which give it stability are a high shear strength, a low permeability and low water absorption.

It is often possible to improve these properties of the soil *in-situ* by various processes such as compaction or stabilization.

Compaction

Compaction is the process of packing together the soil particles by reducing the air voids. This is normally done by mechanical means such as rolling, ramming or vibration. (It should not be confused with the process of consolidation, which is the process of expulsion of water from the voids by continuous pressure.)

To decide if compaction is worthwhile, it is necessary to know:

(1) the natural state of compaction of the soil,
(2) the maximum compaction possible,
(3) the proportion of the maximum compaction which can be obtained on the site.

The state of compaction of the soil is measured quantitatively in terms of the dry density and varies with the moisture content.

Worked examples

8.1 Optimum water content of a soil sample using the standard compaction test

Describe the standard compaction test, stating its object.

In a standard compaction test on a soil ($G_s = 2 \cdot 70$), the following results were obtained:

Water content %	Bulk density (Mg/m³)
5	1·89
8	2·13
10	2·20
12	2·21
15	2·16
20	2·08

Show these results plotted as dry density against water content. On the same axes show the zero air voids (saturation) line for the soil.

What are the values of void ratio, porosity and degree of saturation for the soil at its condition of optimum water content?

Solution The apparatus used consists of a cylindrical metal mould 105 mm dia × 115·5 mm high giving a volume of 1 000 ml. It has a detachable collar and base and is known as a Proctor mould after the name of its developer (Fig. 8.1).

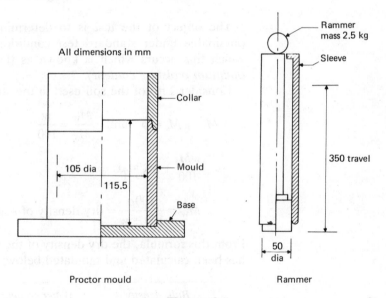

All dimensions in mm

Collar

105 dia

115.5

Mould

Base

Proctor mould

Rammer mass 2.5 kg

Sleeve

350 travel

50 dia

Rammer

Figure 8.1

Soil is compacted in the mould with a metal rammer with a 50 mm diameter face and a mass of 2·5 kg. The height of the fall of the rammer on the soil is 350 mm.

The soil is air-dried and passed through a 20 mm sieve, it is then mixed with a small amount of water and compacted into the mould in three equal layers by giving each layer 27 blows from the rammer. The soil is trimmed to the top of the mould and weighed to determine the wet bulk density. The water content is found and the dry density calculated.

The procedure is repeated at several increasing water contents and a compaction curve of dry density/water content plotted, which should be of the form shown in Fig. 8.2.

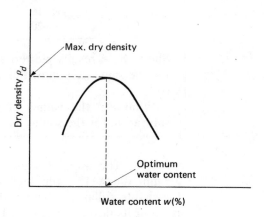

Figure 8.2

The object of the test is to determine the maximum dry density obtainable under standard test conditions and the water content at which this occurs which is known as the *optimum water content* (or *optimum moisture content*).

Consider 1 m³ of the soil used in the standard compaction test.

$$M_W + M_s = \rho \quad \text{and} \quad \frac{M_W}{M_S} = \frac{w}{100}$$

$$\therefore \quad \frac{wM_S}{100} + M_S = \rho$$

$$\therefore \quad M_W = \frac{100\rho}{100 + w} = \text{dry density of soil } \rho_d \qquad (8.1)$$

From this formula, the dry density of the soil at various water contents has been calculated and tabulated below:

Bulk density ρ (Mg/m³)	Water content w (%)	Dry density ρ_d (Mg/m³)
1·89	5	1·80
2·13	8	1·97
2·20	10	2·00
2·21	12	1·97
2·16	15	1·88
2·08	20	1·73

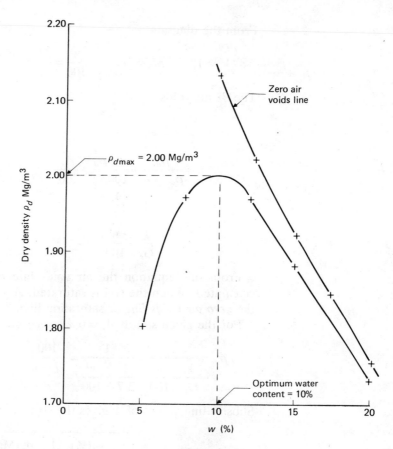

Figure 8.3

The values of ρ_d are plotted against w on Fig. 8.3. To find the zero air voids line, consider the soil-phase diagram shown in Fig. 8.4.

$$V = V_S + V_W + V_A$$

$$\therefore \ V = \frac{M_S}{\rho_w \cdot G_s} + \frac{M_W}{\rho_w} + V_A$$

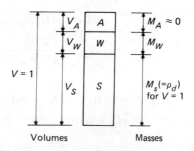

Figure 8.4

From the diagram

$$V = 1, \qquad M_S = \rho_d, \qquad \frac{w}{100} = \frac{M_w}{M_S}$$

The % air voids

$$A_r = \frac{V_A}{V} \times 100$$

$$\therefore \ 1 = \frac{\rho_d}{\rho_w \cdot G_s} + \frac{w \cdot \rho_d}{100\rho_w} + \frac{A_r}{100}$$

$$\therefore \ \rho_d = \rho_w \ \frac{1 - \dfrac{A_r}{100}}{\dfrac{1}{G_s} + \dfrac{w}{100}} \qquad\qquad (8.2)$$

From this equation the air voids line for any value of A_r can be calculated. When the soil is saturated, $A_r = 0$ and the line is known as the *zero air voids line* or saturation line.

For the given soil, if $A_r = 0\%$, since $\rho_w = 1{\cdot}0 \, \text{Mg/m}^3$

$$\rho_d = \frac{1}{\dfrac{1}{G_s} + \dfrac{w}{100}} = \frac{1}{\dfrac{1}{2{\cdot}7} + \dfrac{w}{100}} = \frac{100}{37 + w}$$

Substituting values of w gives the corresponding values of:

w (%)	ρ_d (Mg/m³)
10	2·13
12·5	2·02
15	1·92
17·5	1·84
20	1·75

Using these values, the zero air voids line has been plotted on Fig. 8.3 from which $\rho_{d\,max} = \mathbf{2{\cdot}00 \, Mg/m^3}$ and the optimum water content = **10%**.

Consider 1 m³ of the soil in this condition.

$$M_S = 2{\cdot}00 \, \text{Mg/m}^3$$

$$\therefore \ V_S = \frac{2{\cdot}00}{1{\cdot}00 \times 2{\cdot}7} = 0.74 \, \text{m}^3$$

$$\frac{w}{100} = \frac{M_W}{M_S} = \frac{10}{100}$$

$$\therefore \ M_W = 0{\cdot}10 \times 2{\cdot}00 = 0{\cdot}2 \, \text{Mg}$$

$$\therefore \ V_W = \frac{0{\cdot}2}{1{\cdot}0} = 0{\cdot}2 \, \text{m}^3$$

$$\therefore \; V_A = V - V_W - V_s = 1 - 0.2 - 0.79 = 0.06 \text{ m}^3$$

$$\therefore \; e = \frac{V_V}{V_S} = \frac{V_A + V_W}{V_S} = \frac{0.26}{0.74} = \mathbf{0.35}$$

$$\therefore \; n = \frac{V_V}{V} = \frac{0.26}{1.00} = \mathbf{0.26}$$

$$\therefore \; S_r = \frac{V_W}{V_V} = \frac{0.20}{0.26} = \mathbf{0.77}$$

8.2 Effect of increased compaction on a granular soil

What effect has increased compaction on the properties of a granular soil?

The results of a compaction test on a soil to simulate field conditions were:

Water content w (%)	6.8	8.5	9.4	10.2	11.3	12.5	13.6
Bulk density (Mg/m³)	2.07	2.14	2.18	2.21	2.23	2.21	2.19

Plot the dry density/water content graph and on the same axes, the zero air voids and 5% air voids lines. The specific gravity of the soil particles is 2.7.

If at 10% water content, a heavier roller was available, should it be used on the soil?

Solution For a particular compactive effort, it has been found that when a soil has a low water content, it is not possible to reduce the air voids very much since there is insufficient water in the voids to act as a lubricant enabling the particles to pack closer together. Thus the dry density is comparatively low.

If the water content is high, it is possible to reduce the few air voids but, because of the presence of a large amount of water in the voids, the dry density will again be low.

Between these two extremes, there is a water content known as the optimum moisture content at which the dry density reaches its maximum. This is not a fixed figure however but depends on the amount of compactive effort applied. Increasing the compactive effort has the effect of reducing the value of the optimum water content and increasing the value of the dry density (Fig. 8.5).

The laboratory test can be used to predict the optimum water content and dry density in the field. It is important therefore that the energy input used is equivalent to that likely to be applied by the compaction plant. The standard test has been outlined in question 8.1 but for heavier compaction equipment, a modified test is used. The procedure is similar but a 4.5 kg hammer is used to compact five layers

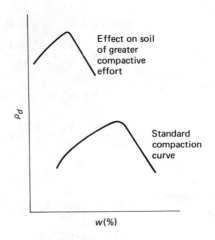

Effect on soil of greater compactive effort

Standard compaction curve

ρ_d

$w(\%)$

Figure 8.5

of soil with a free fall of 450 mm. For the given results:

The dry density is calculated using eqn (8.1) $\rho_d = \dfrac{100\rho}{100 + w}$

w (%)	6·8	8·5	9·4	10·2	11·3	12·5	13·6
ρ_d (Mg/m³)	1·94	1·97	1·99	2·00	2·00	1·96	1·93

The zero and 5% air voids lines are calculated using eqn (8.2).

w (%)	10	11	12	13	14
$A_r = 0$	2·13	2·08	2·04	2·00	1·96
$A_r = 5\%$	2·02	1·98	1·94	1·90	1·86

These values are plotted on Fig. 8.6.

The effect of increasing the compacting force is to shift the curve upwards as shown on Fig. 8.6. At 10% water content on the initial curve, the dry density is 2·00 Mg/m³ and the optimum water content has not been reached.

On the second curve, the dry density corresponding to 10% water content is 2·05 Mg/m³ but it is beyond the optimum water content.

If compaction continues beyond the optimum water content, the soil will be softer than if it had stopped before that content is reached. In this case therefore, it would be inadvisable to use the heavier roller.

Stabilization

Stabilization is the process of improving the soil properties by adding something to the soil.

A mechanically stable soil usually consists of a mixture of coarse and fine aggregate with a silt and clay binder in correct proportions. The

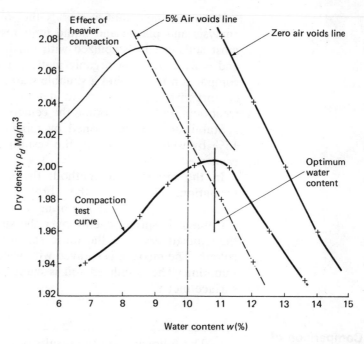

Figure 8.6

process of mechanical stabilization is the improvement of a soil by the addition of the particle sizes which are missing from its natural grading.

8.3 Methods of stabilizing soils

> What are the objects of soil stabilization? State the principal materials used for this purpose.
>
> Give a short account of the stabilization of soil by Portland cement. Include a brief summary of the procedure in the field using the mix-in-place method.   (ICE)

Solution A soil which has been well compacted at a suitable water content will usually have a satisfactory load bearing capacity. If, however, the water content is increased, there is a danger of the soil becoming unstable. The objects of soil stabilization are therefore:

(1) to increase the stability of the soil;

(2) to maintain the soil in a state of high stability.

The principal materials used in addition to those mentioned in solution (8.3) are:

(*a*) Portland cement,

(*b*) bituminous materials.

In order to assess the suitability of a soil for stabilization preliminary tests are carried out in the laboratory.

It is most important to classify the soil. This is done by particle size analysis and plastic and liquid limit tests (see chapter 1). In general most soils can be stabilized with Portland cement except heavy clays and soils containing organic matter. The other tests necessary are a compaction test to find a suitable water content and dry density and a compressive strength test to determine the suitability of a soil for treatment and to determine the cement content required.

Stabilization is performed by mixing the soil with between 5 and 15% by weight of cement; the resulting mixture is referred to as *soil cement*.

In the mix-in-place method, the site is levelled to the required formation. The soil is then broken up to the required depth of treatment and pulverized to bring it to a fine tilth.

Cement is spread evenly over the surface of the prepared soil and then mixed dry into the loose soil until the mixture has a uniform colour. The mixture is sprayed with water and compacted by rolling or ramming. The stabilized soil is cured for seven days by keeping the surface moist.

8.4 Comparison of optimum water content obtainable in the laboratory and in the field

The following are the results of a standard compaction test on a sand/cement mixture having an equivalent grain specific gravity of 2·70:

Water content (%)	5	8	10	12·5	16	20
Compacted dry density (Mg/m³)	1·64	1·78	1·85	1·89	1·84	1·73

Plot these results and on the same axes plot the zero air voids line. What percentage of air voids ($A_r\%$) exists in the sample at optimum water content?

Experience shows that in field actual air voids may be $(A_r \pm 2\cdot5)\%$ and the water content may range between optimum plus 3% and optimum minus 4%. On the same axes as before plot the zone into which the field product will fall, marking clearly the values of dry density at the extremes.

What field control tests will be required? (ICE)

Solution The given values of dry density ρ_d and water content w are plotted on Fig. 8.7.

From the figure, the optimum water content $= 12\cdot5\%$ and the corresponding dry density, $\rho_{d\,max} = 1\cdot89\,\mathrm{Mg/m^3}$.

The zero air voids line is calculated using eqn (8.2). With $A_r = 0$ and $G_s = 2\cdot7$,

$$\rho_d = \frac{1}{\dfrac{1}{2\cdot7} + \dfrac{w}{100}} = \frac{100}{37 + w}$$

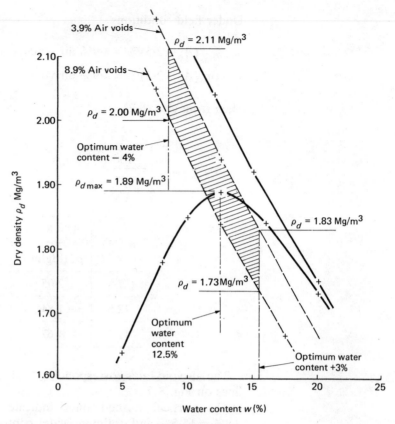

Figure 8.7

$$\therefore \text{ when } w = 20\%, \; \rho_d = 1 \cdot 75 \text{ Mg/m}^3$$
$$w = 15\%, \; \rho_d = 1 \cdot 92 \text{ Mg/m}^3$$
$$w = 12\%, \; \rho_d \doteq 2 \cdot 04 \text{ Mg/m}^3$$

Using these values, the zero air voids line can be added to the figure. Considering 1 m³ of soil in this condition

$$V_S = \frac{1 \cdot 89}{2 \cdot 7 \times 1 \cdot 0} = 0 \cdot 7 \text{ m}^3$$

$$w = \frac{M_W}{M_S} \times 100$$

$$\therefore M_W = \frac{12 \cdot 5 \times 1 \cdot 89}{100} = 0 \cdot 236 \text{ Mg}$$

$$\therefore V_W = \frac{0 \cdot 236}{1 \cdot 0} = 0 \cdot 236 \text{ m}^3$$

$$\therefore V_A = 1 - 0 \cdot 236 - 0 \cdot 70 = 0 \cdot 064 \text{ m}^3$$

$$\therefore A_r = \frac{0 \cdot 064}{1 \cdot 0} \times 100 = \textbf{6·4\%}$$

Under field conditions

$$(A_r + 2 \cdot 5)\% = 8 \cdot 9\% \text{ air voids}$$

$$\text{or } (A_r - 2 \cdot 5)\% = 3 \cdot 9\% \text{ air voids.}$$

Then from eqn (8.2)

$$\rho_d = \rho_w \frac{1 - \dfrac{A_r}{100}}{\dfrac{1}{G_s} + \dfrac{w}{100}}$$

w (%)	$Ar = 8 \cdot 9\%$	$A_r = 3 \cdot 9\%$
	ρ_d (Mg/m³)	ρ_d (Mg/m³)
7·5	2·05	2·16
12·5	1·84	1·94
17·5	1·67	1·76

The air void lines corresponding to these values are shown in dashed lines on Fig. 8.7.

The vertical dashed lines indicate optimum water content at $+3\% = 15 \cdot 5\%$ and optimum water content at $-4\% = 8 \cdot 5\%$.

The shaded portion of Fig. 8.7 then represents the zone into which the field product may fall and the corresponding values of dry density have been indicated.

The field control tests required are:

(1) Determination of water content (by pycnometer method: chapter 1).
(2) Compressive strength tests on samples of the soil compacted into cylindrical moulds.
(3) Determination of dry density of soil-cement *in situ* (by sand replacement method: chapter 1).
(4) Determination of cement content by chemical means.

Design of road and airfield runway pavements

The *pavement* of a road or airfield runway is defined as the whole of the artificial construction made to support traffic above the subgrade.

The *subgrade* is the natural foundation which receives the load from the pavement.

The elements which compose a pavement are shown on Fig. 8.8 but it should be noted that not all the elements are necessarily present in a particular pavement.

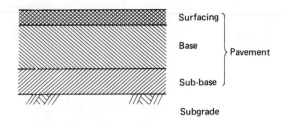

Surfacing ⎫
Base ⎬ Pavement
Sub-base ⎭
Subgrade

Figure 8.8

8.5 Test procedure for measuring the California Bearing Ratio of a soil

Describe briefly the test procedure for measuring the California Bearing Ratio (CBR) of a soil.

The following results were obtained from a CBR test on a compacted sample of sandy clay:

Penetration of plunger (mm)	Load on plunger (kN)	Penetration of plunger (mm)	Load on plunger (kN)
0	0·05	4·00	2·04
0·25	0·11	4·25	2·13
0·50	0·25	4·50	2·19
0·75	0·38	4·75	2·23
1·00	0·56	5·00	2·26
1·25	0·78	5·25	2·33
1·50	0·99	5·50	2·36
1·75	1·14	5·75	2·43
2·00	1·34	6·00	2·45
2·25	1·44	6·25	2·52
2·50	1·51	6·50	2·54
2·75	1·65	6·75	2·56
3·00	1·74	7·00	2·60
3·25	1·85	7·25	2·62
3·50	1·93	7·50	2·64
3·75	1·96		

Plot the load/penetration curve and estimate the CBR of the soil if the standard values at 2·5 mm and 5·00 mm are 13·24 kN and 19·96 kN respectively.

Solution The apparatus used in the laboratory consists of a metal mould 152 mm internal diameter and 127 mm high with a detachable base and collar 50 mm high which can be attached to the mould during the preparation of the specimen; a loading frame capable of applying a load of up to 30 kN smoothly to a cylindrical plunger having a diameter of 49·6 mm (Fig. 8.9).

The soil to be tested is brought to a moisture content of approximately that expected on the site and broken into lumps less than

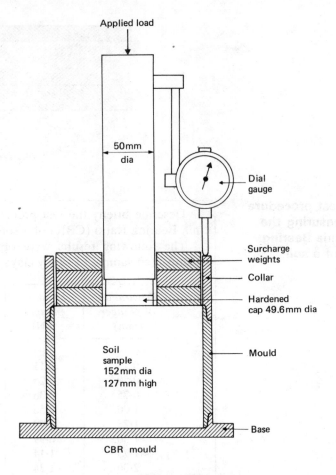

Applied load

50 mm
dia

Dial
gauge

Surcharge
weights

Collar

Hardened
cap 49.6 mm dia

Soil
sample
152 mm dia
127 mm high

Mould

Base

CBR mould

Figure 8.9

20 mm dia. Sufficient of the wet soil is tamped continuously into the mould with base and collar attached.*

After tamping, a filter paper is placed on top of the soil, followed by a 50 mm thick displacer disc and the specimen is compressed in a compression machine until the disc is flush with the top of the collar. The displacer disc and collar are then removed and the mould and soil weighed to determine the mass of soil and hence the wet density.

Annular surcharge weights may be placed on the surface of the

---

* The amount of wet soil required to give the necessary state of compaction is calculated from the formula:

$$\text{Mass of wet soil required (g)} = \frac{23 \cdot 2 \left(1 - \dfrac{A_r}{100}\right)(100 + w)}{\dfrac{1}{G_s} + \dfrac{w}{100}}$$

where A_r = air content %
w = water content %.

specimen to simulate the weight of construction above the soil being tested.

The plunger is seated on the sample under a load of 0·05 kN (which is neglected in the calculations) and the load and penetration gauges set to zero.

The load is then applied to the plunger so that the rate of penetration is constant at 1·0 mm min. The load on the plunger is recorded at penetrations for each 0·25 mm up to 7·5 mm.

After the completion of the loading, the moisture content of a sample taken from the soil is found. The test results are plotted in the form of a load/penetration curve. If the curve is initially concave upwards it has to be corrected by drawing a tangent to the curve at its steepest slope and the penetration scale is adjusted to the point where this tangent cuts the axis.

From the curve, the corrected loads to produce a penetration of 2·5 mm and 5·0 mm are read off and expressed as percentages of the standard loads for crushed stone of 13·24 kN and 19·96 kN respectively. The higher of the two percentages is known as the CBR for the soil.

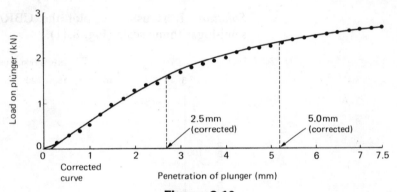

Figure 8.10

The given test results are plotted on Fig. 8.10 (the initial 0·05 kN having been subtracted from the given values).

It can be seen that the initial part of the curve required correction and from the corrected scale:

$$\text{Load at } 2\cdot5 \text{ mm} = 1\cdot6 \text{ kN}$$

$$\therefore \text{ CBR} = \frac{1\cdot6}{13\cdot24} \times 100 = 12\%$$

$$\text{Load at } 5\cdot00 \text{ mm} = 2\cdot3 \text{ kN}$$

$$\therefore \text{ CBR} = \frac{2\cdot3}{19\cdot96} \times 100 = 11\cdot5\%$$

The higher value is taken.

$$\therefore \text{ CBR for soil} = \mathbf{12\%}$$

8.6 Thickness of pavement elements using results of CBR tests

Using the data given below, determine the total thickness of a flexible road pavement and the thickness of each structural element, illustrating your answer with a sketch.

	CBR (%)
Sandy clay subgrade	9
Sand sub-base	22
Well-graded gravel base	80

Design data for relevant loading conditions:

CBR (%)	Thickness of construction (mm)
5	375
10	285
15	225
20	190
30	150
80	100

Solution It is usual to plot the CBR/depth of construction to a semi-logarithmic scale (Fig. 8.11).

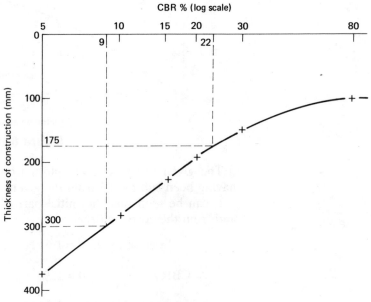

Figure 8.11

From the information given:

CBR of subgrade = 9%

∴ Total thickness of construction above subgrade = 300 mm (Fig. 8.11).

CBR of sub-base = 22%

∴ Thickness of construction required above sub-base = 175 mm

CBR of base = 80

∴ Thickness of construction above base = 100 mm.

With this information the various thicknesses of the structural elements of the pavement can be determined.

Total thickness of pavement = 300 mm

Thickness above sub-base = thickness of base + surfacing

= 175 mm

Thickness above base = thickness of surfacing = 100 mm

∴ Thickness of base = 175 − 100 = 75 mm

∴ Thickness of sub-base = 300 − 175 = 125 mm.

A sketch of the construction is shown on Fig. 8.12.

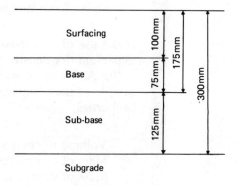

Figure 8.12

Problems

1. Derive, from first principles, an expression for the percentage of air voids in a partially saturated soil in terms of ρ_d, G_s and w.

A sample of clay is compacted at a water content of 20% to a bulk density of $1 \cdot 84$ Mg/m^3, $G_s = 2 \cdot 74$. Determine the percentage of air voids. It is found that if the soil is compacted in the same way with a water content of 25% the bulk density is unaltered; how are the dry density and percentage of air voids affected?

(ULKC)

$$\left(A_r = \left[1 - \frac{\rho_d}{\rho_w} \left(\frac{1}{G_s} + \frac{w}{100} \right) \right] \times 100; \ \rho_d = 1 \cdot 53 \text{ Mg/m}^3, \ A_r = 13 \cdot 5\%; \right.$$

$$\left. \rho_d = 1 \cdot 47 \text{ Mg/m}^3, \ A_r = 9 \cdot 5\% \right)$$

2. (*a*) When a standard compaction test was performed on samples of a heavy clay at four different moisture contents the following dry densities were obtained:

Sample number	1	2	3	4
Water content (%)	20	21·2	24·8	26
Dry density (Mg/m³)	1·57	1·59	1·58	1·56

To obtain another point on the dry density/water content curve another sample was subjected to the same compactive effort and the following data obtained:

Mass of mould + compacted wet soil	7 665 g
Mass of mould	5 705 g
Capacity of mould	1 000 ml
Mass of sub-sample taken from mould	202 g
Mass of sub-sample after thorough drying	165 g

Make use of the above-mentioned information to determine the maximum dry density and the optimum water content.

(*b*) At a site on which the soil tested above is being rolled, a sample of the compacted material is found to have the following properties:

Volume of compacted sample	1 348 ml
Mass of compacted sample	2.16 kg
Mass of sample after thorough drying	1·73 kg

Determine
  (i) the relative compaction (ratio of *in-situ* density to maximum dry density) being achieved by rolling on the site.
  (ii) the percentage of air voids in the rolled material.
Specific gravity of soil particles, $G_s = 2.70$.

(SCOTEC)

[(*a*) 1·63 Mg/m³, 22·5%; (*b*) 0·98, 50%]

3. Use the results of a field compaction test which are given below to plot the variation of dry density with water content. Include lines representing 100% and 80% saturation on the same graph and show separately how the positions of these lines were calculated. The specific gravity of the soil grains is 2·70.

Sample number	1	2	3	4	5	6
Compacted volume ($m^3 \times 10^{-3}$)	0·82	0·78	0·74	0·75	0·73	0·81
Mass when wet (kg)	0·157	0·168	0·163	0·161	0·158	0·169
Mass when dry (kg)	0·146	0·152	0·146	0·142	0·141	0·145

When the specification for compaction on a site is being decided there are a number of important points to consider such as the local weather conditions. List three points you consider important. (CEI)

4. 100 kN of a soil which has a bulk density of 1·60 Mg/m^3 have been excavated and the hole refilled with a different soil having a grain specific gravity of 2·66. The results of a standard laboratory compaction test on this fill are given below.

Dry density (Mg/m^3)	1·67	1·76	1·79	1·77	1·72	1·61
Water content (%)	11	13	15	17	19	21

If the compaction of the fill is to achieve 95% of the maximum placement dry density in the standard test, what is the maximum degree of saturation which can be permitted and how many kN of fill will be needed at this degree of saturation?

Discuss the techniques of quality control for field compaction including the part played by a standard laboratory test. (CEI)

($S_r = 54\%$, 118·4 kN)

5. The following results were obtained from a CBR test on a compacted specimen of sandy clay:

Penetration (mm)	Load (kN)	Penetration (mm)	Load (kN)
0·25	0·08	4·00	2·40
0·50	0·11	4·25	2·59
0·75	0·19	4·50	2·71
1·00	0·24	4·75	2·86
1·25	0·30	5·00	3·00
1·50	0·48	5·25	3·10
1·75	0·60	5·50	3·21
2·00	0·79	5·75	3·33
2·25	0·99	6·00	3·42
2·50	1·20	6·25	3·50
2·75	1·42	6·50	3·60
3·00	1·61	6·75	3·70
3·25	1·82	7·00	3·77
3·50	2·01	7·25	3·82
3·75	2·20	7·50	3·92

Plot the load/penetration curve and estimate the CBR of the soil. The standard values for crushed stone are 13·24 kN for 2·5 mm penetration and 19·96 kN for 5·0 mm.

Explain how the thickness of a flexible road pavement can be determined from the CBR of the subgrade.

(CBR = 17% approx.)

6. Describe briefly the CBR test and explain the principles of its use in pavement design.

A road pavement is to consist of a bituminous surface on a crushed stone base. The sub-base consists of a poorly graded gravel having a CBR (compacted) of 12% and the subgrade has a CBR of 3·5%. Using the design data given below find (*a*) the minimum combined thickness of surfacing and base; (*b*) the minimum thickness of sub-base.

CBR (%)	Thickness of pavement for medium traffic (mm)
2	600
3	500
5	375
10	250
15	200
20	175

(Surface + base = 225 mm; sub-base = 225 mm)   (ICE)

7. Describe concisely the California Bearing Ratio test and explain the conditions for which it is most suited. Illustrate your answer with neat sketches.

Readings as shown in the table are taken during a CBR test on a sample of compacted sub-grade.

Penetration (mm)	Measured load (kN)	CBR standard load (kN)
0·625	0·15	—
1·250	0·18	—
1·875	0·40	—
2·500	0·73	13·50
3·750	1·32	—
5·000	1·80	20·00
7·500	2·34	25·50
10·000	2·53	31·00
12·50	2·91	35·00

Calculate the CBR value of the soil on the basis of the CBR standard values given in the third column.

CBR design charts show the following relationship for the anticipated traffic:

CBR value (%)	Minimum depth of construction required (mm)
5	473
10	318
20	203
40	140
80	89

Design an appropriate pavement assuming that materials of 15%, 30%, and 45% CBR are available.　　　　(ULKC)

(10%; 160 mm sub-base 320 mm; total thickness)

Index